Blast and Ballistic Loading of Structures

To
Elizabeth, Caroline and Rachel
and
Janice, Alexander, Iona, Douglas, Amy and Alastair

Blast and Ballistic Loading of Structures

P. D. Smith
Senior Lecturer, Civil Engineering Group,
Cranfield University, Royal Military College
of Science

J. G. Hetherington
Head, Design Group, Cranfield University,
Royal Military College of Science

 Routledge
Taylor & Francis Group

LONDON AND NEW YORK

First published 1994 by Butterworth-Heinemann

2 Park Square, Milton Park, Abingdon, Oxfordshire OX14 4RN
52 Vanderbilt Avenue, New York, NY 10017

Routledge is an imprint of the Taylor & Francis Group, an informa business

First issued in paperback 2019

British Library Cataloguing in Publication Data
A catalogue record for this book is available from the British Library

Library of Congress Cataloguing in Publication Data
A catalogue record for this book is available from the Library of Congress

ISBN 978-0-7506-2024-6 (hbk)
ISBN 978-0-367-86687-7 (pbk)

Contents

Preface

The authors are members of the academic staff at the Royal Military College of Science and have been instrumental in establishing an active group comprising lecturing and research staff concerned with the response of structures to blast and ballistic loading.

The group has developed a taught Master's programme entitled 'Weapons Effects on Structures' which has attracted students from all over the world. In addition a number of specialist short courses have been presented in the United Kingdom, the United States of America and Australia. As worldwide interest has developed in this subject area the group has undertaken an increasing number of research programmes for a range of British and overseas sponsors. These programmes have included: studies into the ballistic performance of composite armours; the attenuation of blast resultants by labyrinthine entranceways; the determination of resistance-deflection functions for wall panels with openings; the development of anti-personnel mine protective footwear; the measurement of leakage pressures behind blast walls; the penetration of oil-well shaped charges into targets simulating down-hole conditions; damage simulation experiments aimed at improving safety on offshore structures and methods of attenuating dynamic structural loading from underwater explosions. In addition the group has acted in a consultancy capacity giving advice on a number of projects in which there is a potential terrorist threat or the chance of an accident producing abnormal structural loads.

It is against this background that this book has been written. The authors have drawn upon the research and experience of their colleagues and would like to express their gratitude for their advice and assistance which has been so generously given. We are grateful to the college for support in the preparation of this text. In particular we would like to thank Mrs Jean Mosley for her skill in producing the diagrams and to Mrs Ros Gibson and Mrs Jan Price for their help in manuscript preparation.

P. D. Smith
J. G. Hetherington

1 Introduction

1.1 Basic concepts of protection

The aim of this book is to help you to design structures with a specified level of *robustness*. By robustness we mean the ability of a structure to survive a certain level of blast or ballistic loading. The techniques which we describe, of course, can be used to assess the robustness of existing structures and so, inevitably, to determine how best to destroy them. Our purpose, however, is to help the designer to provide a structural environment in which life and property are protected against hostility and man-made disasters.

The first stage is to devise ways of preventing the attack or the accident happening at all, or at least making sure it happens at a safe distance. The blast effects of a bomb diminish dramatically with range. Keeping vehicle bombs away from your structure is probably the single, most cost-effective device you can employ.

The provision of structural protection can be very expensive and often impedes the activities which the structure is designed to accommodate. For this reason it is rarely feasible or desirable to make any structure totally immune to every conceivable incident.

1.2 Protective design

An essential preliminary step in protective design is to survey the spectrum of potential incidents. We then eliminate those which fall into one or more of the following categories.

1 The chance of the incident happening is negligible.
2 The effects of the incident are negligible.
3 Protection against the effects of the incident would be prohibitively expensive.
4 Protection against the effects of the incident would obstruct the proper functioning of the structure.

Tables 1.1, 1.2 and 1.3 summarise the principal threats to which a structure may be subjected.

Table 1.1 *Explosive threats*

Threat	Description
Nuclear device	Bomb releasing massive quantities of energy, causing thermal, electromagnetic radiation and blast effects (only blast effects within the scope of this work).
Gas and vapour cloud explosion	Military fuel/air munitions produce a detonable mixture generating blast loading over a large target area. Accidental explosions, either deflagrative or detonative, occur in the process industries and in the domestic environment.
Dust explosions	Accidental deflagrations occur in such locations as flour mills, grain silos, coal mines etc. and can cause extensive blast damage.
Uncased high explosive	In-contact charges used for breaching structures. At small stand-offs, high intensity localised blast loads are generated.
High explosive bombs	Manufactured munitions, normally delivered aerially, including: *General purpose bombs* which cause damage by blast and fragmentation. *Light case bombs* which primarily produce blast damage. *Fragmentation bombs* which are effective against personnel and light equipment, and *Armour piercing bombs* which are designed to penetrate protected structures.
Artillery shells	Fragmenting munitions delivered by heavy guns which, depending on fuzing, could penetrate concrete, soil etc. before detonation.
Vehicle bomb	Common terrorist weapon designed to cause blast damage to structures.
Incendiary bomb	A bomb in which blast effects are augmented by a fireball from a burning fuel such as petrol.
Package bomb	Common hand-portable terrorist device infiltrated inside a structure to cause a high level of damage.
Missiles and rocket propelled munitions	Bombs and shaped-charge devices, the delivery of which is achieved by, or assisted by, rocket propulsion.
Mortars	Military mortars cause similar damage to that caused by artillery shells, although range is less and accuracy inferior. The improvised mortar, a potent terrorist weapon, is designed to overfly perimeter security, penetrate structures and detonate internally.
High explosive squash head (HESH) rounds	An anti-tank round, containing high explosives, designed to detonate on impact and cause scabbing at the target. Also effective against structural targets.

Table 1.2 *Shaped-charge threats*

Shaped-charge devices	A generic term for rounds in which explosive is contained behind a metal liner. When detonated, the explosive forms the liner into a highly penetrative jet or slug.
High explosive anti-tank rounds (HEAT)	A large calibre shaped-charge round with conical liner, fired from a tank gun and designed to penetrate tank armour.
Rocket propelled grenades (RPG)	A medium calibre, rocket propelled shaped-charge round fired from a hand-held launcher.
Bomblets	Aerially delivered, small-calibre, shaped-charge devices for top attack of vehicles.
Self-forging fragments	Shaped-charge devices using a dished plate in place of conical liner. Used in off-route mine applications.
Linear shaped charges	Cutting charges employing a V shaped liner used in demolition.

Table 1.3 *Other ballistic threats*

Small arms ammunition	Inert anti-personnel projectiles.
Armour piercing rounds	Inert projectiles of any calibre (from small arms to tank ammunition) designed to penetrate armour.
Sabot-mounted rounds	Sub-calibre rounds (e.g. long rod penetrators) designed to penetrate tank armour.
Fragments	Inert, often irregular, projectiles of widely varying mass and velocity, generated from fragmenting bomb cases etc.

Before proceeding to a detailed examination of the effects caused by these incidents, one or two general observations can be made.

1.2.1 Materials of construction

In order to withstand the transient loads generated by any one of these threats, the elements of a structure need to be both massive and able to absorb large amounts of energy. For this reason nearly all purpose-built protective structures are constructed of reinforced concrete.

1.2.2 Vehicle and package bombs

Security procedures at the entrance to a facility, in conjunction with effective fencing and boundary surveillance, will greatly reduce the possibility of a vehicle or package bomb being infiltrated. There are many situations, however, where such a visible display of officialdom is undesirable. Since even a relatively small package bomb can cause devastation if it is detonated inside a building, it is advisable to isolate those parts of the structure which are accessible to the public from the more sensitive or critical areas of the building.

1.2.3 The ballistic threat

The ballistic threat comprises those threats which are designed to damage the target by penetration rather than blast (Tables 1.2 and 1.3). Although most forms of ballistic attack cause only minor structural damage they constitute a potent threat for personnel and equipment. Obscuration of the target, by means of trees, tinted glass etc., will greatly reduce the effectiveness of such an attack.

1.3 Detailed examination of the effects – the scope of this work

When a bomb is detonated, very high pressures are generated which then propagate away from the source. The process by which this pressure pulse is generated, in both nuclear and chemical devices, is described in Chapter 2. The propagation of the wave through air, water, solids and soil is described in Chapters 3, 5, 6 and 7, while the special condition of internal blast loading is examined in Chapter 4. When the wave strikes the target, a transient load is applied to the structure. Chapters 8, 9, 10 and 11 describe the techniques which are available for predicting the response of the structure to this transient loading.

The ability of a structure to offer protection against a direct fire ballistic attack derives from the energy absorbing capacity of the fabric of the structure. Mechanisms of ballistic penetration and penetration prediction techniques are discussed in Chapter 12.

One of the most effective methods of providing protection for a structure is to bury it. The analysis of buried structures has conventionally been based on empirical rules relating to specific geometries and loading configurations. Chapter 13 describes a powerful method of analysis, based on the theory of engineering plasticity, which is of wide applicability.

Chapter 14 draws on the ideas developed in the earlier chapters to formulate general design principles and advice for protective structures. Reference is made to the various design guides and manuals which are available to assist the designer. Worked examples are included in the text to assist the reader in applying the various analytical techniques which have been discussed. A more comprehensive and wide-ranging package of worked examples is presented in Chapter 15.

Each chapter ends with a list of references and, where appropriate, a bibliography. A list of symbols used in each chapter is also presented.

Historically, the various topics in the book have developed their own notations independently with which the authors have tried to keep faith. For this reason, a particular symbol will have a unique meaning within a specific chapter but may have an alternative meaning elsewhere in the text.

2 Introduction to explosives

2.1 Explosions

Explosions can be categorised as physical, nuclear or chemical events. Examples of *physical explosions* include:

1 the catastrophic failure of a cylinder of compressed gas or another type of pressure vessel. The failure of such vessels is accompanied by blast waves and the generation of high velocity debris;
2 the eruption of a volcano. A dramatic example is provided by the island of Krakatoa in the Strait of Sunda, Indonesia, the greater part of which was destroyed in 1883;
3 the violent mixing of two liquids at different temperatures or the mixing of a hot particulate material with a cool liquid. In latter examples there is no chemical reaction: the explosion results from the rapid conversion to vapour of the cooler fluid at a rate that does not permit venting, resulting in shock wave generation.

In a *nuclear explosion* the energy released arises from the formation of different atomic nuclei by the redistribution of the protons and neutrons within the interacting nuclei. Thus, what is sometimes described as atomic energy is really nuclear energy since it arises from particular nuclear reactions. The two kinds of nuclear interactions that produce large amounts of energy in a short time are the processes of 'fission', or splitting of (heavy) atoms such as those of certain isotopes of uranium, and 'fusion', or joining together of (light) atoms such as the hydrogen isotope deuterium.

Because the forces that exist between the components of an atomic nucleus are orders of magnitude greater than those between atoms, the energy associated with a nuclear reaction will be many times greater than that released from the same mass of material in a 'conventional' or chemical reaction. The bombs dropped on Hiroshima and Nagasaki in the Second World War had yields expressed in kilotonnes of TNT of 12.5 and 22 respectively. The release of nuclear energy can be divided into three categories: kinetic energy, internal energy and thermal energy. The first two types comprise approximately half of the energy released and result in the formation of blast waves (just as in the case of chemical explosives) while most of the remainder is delivered as heat and thermal radiation.

A *chemical explosion* involves the rapid oxidation of fuel elements (carbon and hydrogen atoms) contained within the explosive compound. The oxygen needed for this reaction is also contained within the compound – the presence of air is thus not necessary. To be useful, a chemical explosive must only explode when it is required to do so and should, under normal conditions, be inert and stable. The rate of reaction (much greater than the burning of a fuel in atmospheric air) will determine the usefulness of the explosive material for practical applications. Most practical explosives are 'condensed', meaning that they are either solids or liquids.

When the explosive is caused to react it will decompose violently with the evolution of heat and the production of gas. The rapid expansion of the gases results in the generation of shock pressures in any solid material with which the explosive is in contact or blast waves if the expansion occurs in a medium such as air.

2.2 Thermodynamics of explosions

Consider a general fuel containing carbon, hydrogen, oxygen, nitrogen and sulphur. Its complete (or stoichiometric) oxidation reaction can be written:

$$C_aH_bO_cN_dS_e + \left(a + \frac{b}{4} - \frac{c}{2} + e\right)O_2 \rightarrow aCO_2 + \frac{b}{2}H_2O + \frac{d}{2}N_2 + eSO_2 \qquad (2.1)$$

If this reaction is one of combustion (burning), the heat of reaction is called the heat of combustion ΔH_c. It is the enthalpy needed to be added to the system when the reactants at some initial pressure and temperature react to form the products at the same pressure and temperature. In combustion or when an explosive decomposes, the reaction is described as exothermic, that is, heat is produced and as a consequence the change in enthalpy is negative.

Thus, the oxidation of carbon to form carbon dioxide, as described by the reaction

$$C + O_2 \rightarrow CO_2 \qquad (2.2)$$

is accompanied by an enthalpy change of -393 kJ/mol. Similarly for hydrogen we have

$$H_2 + \tfrac{1}{2}O_2 \rightarrow H_2O \qquad (2.3)$$

and an enthalpy change of -242 kJ/mol.

In an explosive compound these and other elements are bonded together. Because the compound decomposes as the reaction takes place, its own heat of formation (which may be positive or negative) will be yielded up. It is thus usual to define the 'heat of explosion', ΔH, sometimes written as Q, as

$$|\Delta H| = Q = |\Sigma(\Delta H(\text{products}) - \Delta H(\text{explosive}))| \qquad (2.4)$$

In this expression the enthalpy change associated with the products will be much larger than that of the explosive so that ΔH will have a large negative value. Accompanying this enthalpy change will be a large entropy change ΔS which, because gaseous products are formed, will be positive.

If these two quantities are combined as Gibbs Free Energy change ΔG (which may be taken as a measure of the work capacity of the explosive) where

$$\Delta G = \Delta H - T\Delta S \qquad (2.5)$$

and T is the temperature of the products, the result will always be large and negative for explosive materials. Indeed, all explosions, be they physical, nuclear or chemical, are accompanied by a large negative change in Gibbs free energy.

2.3   Mixtures and compounds

2.3.1   Explosive mixtures

If the gases hydrogen and oxygen are mixed in the proportions 2:1 by volume the mixture can be readily exploded by a spark or a flame and the violent reaction described by Equation 2.3 will occur accompanied by an enthalpy change of 242 kJ/mol. Since in a stoichiometric mixture of these gases there are 2 grammes of hydrogen and 16 grammes of oxygen this equates to approximately 13.4 kJ per gramme of mixture which is an impressive yield. However, as a practical explosive mixture, hydrogen and oxygen is deficient: as gases at room temperature they are too bulky and as liquids they are very cold and highly volatile and present difficult handling problems. Except for applications in space rocketry, the use of liquid oxygen ceased a long time ago.

The oldest solid explosive mixture is 'black powder' or 'gunpowder' which originated in China some time before AD 1000. It was first described in Europe by Roger Bacon (1214–1292). The basic composition of modern gunpowder is potassium nitrate, charcoal and sulphur in the proportions 15:3:2 which is prepared in such a way so as to provide an intimate mixture to allow as fast a reaction rate as possible. Even then, the reaction rate is rather slow when compared with a modern, condensed high explosive compound, but the material has found many applications as a propellant for rockets and guns, in demolition work and rock-blasting. The decomposition reaction of gunpowder is difficult to describe. A simplified reaction equation is:

$$4KNO_3 + 7C + S \rightarrow 3CO_2 + 3CO + 2N_2 + K_2CO_3 + K_2S \qquad (2.6)$$

Of the products, only about 43% by mass are gaseous, the last two (K_2CO_3 and K_2S) being solids. This gas production is only about a quarter of that produced by a modern, condensed high explosive while the heat generated is about half. As a consequence, gunpowder is not used commercially in any large quantities, having been replaced by more efficient explosive mixtures which find application in, for example, blasting operations in quarrying.

2.3.2 Explosive compounds

(i) Carbon-hydrogen-oxygen-nitrogen compounds

The nineteenth century saw a rapid growth in organic chemistry. One of the most important discoveries of the period (first reported by Pelouze in 1838) was the effect of adding a mixture of concentrated nitric acid and sulphuric acid (this latter to remove water produced in the reaction) to a material containing cellulose such as cotton, linen or wood. The following reaction occurs producing di-nitrocellulose and water:

$$C_6H_{10}O_5 + 2HNO_3 \rightarrow C_6H_8(NO_2)_2O_5 + 2H_2O \qquad (2.7)$$

The resulting compound, containing carbon, hydrogen, oxygen and nitrogen (sometimes described as a CHON compound), contains a larger amount of the elements nitrogen and oxygen than the original material implying that one or more hydrogen atoms has been replaced in the structure. Furthermore, the compound was found to be rather unstable and could easily be ignited: in some cases explosion occurred under the action of a small impact.

The action of these acids was examined by the Swiss Schönbein in 1846, with particular reference to cellulose in the form of cotton wool. His method was to immerse finely carded cotton wool in a mixture of equal measures of the acids for a few minutes. The material was then taken from the acid bath, plunged into cold water and washed till every trace of acid was removed and then dried at a temperature less than 100°C. The resulting material is gun-cotton, often referred to as nitrocellulose, which was seen as a new agent for blasting or as a propellant in fire-arms.

The use of early samples of gun-cotton was fraught with danger because of its instability. In one incident at Hill's factory at Faversham, Kent, in 1847, a spontaneous explosion occurred that was so catastrophic that the company ceased manufacture, going so far as to destroy any remaining stocks of the material by burying them. Eventually improvements in manufacturing techniques led to large-scale manufacture throughout Europe with factories established at Waltham Abbey, Paris and Hirtenberg, Austria.

The action of nitric acid on glycerine is of the same kind as that on cellulose. The discovery was made as the result of work by several European scientists with the final step to successful manufacture being taken by the Italian Sobrero in 1846. The chemical change that is effected is the substitution of three hydrogen atoms by three $-NO_2$ groups as shown below:

$$C_3H_8O_3 + 3HNO_3 \rightarrow C_3H_5(NO_2)_3O_3 + 3H_2O \qquad (2.8)$$

The resulting compound is nitroglycerine, a dense, oily liquid that explodes with great violence under little provocation. The material was originally used for blasting purposes and was known as 'blasting oil' but its inherent instability led to a number of severe accidents.

The safety of the material was improved step by step by Alfred Nobel who, in 1867, managed to stabilise it by incorporating it with a siliceous material, kieselguhr. He marketed it under the trade name of Dynamite.

Its invention heralded the start of a new era because, after centuries of use, black powder could now be succeeded by high explosives. It is worth noting that Nobel was also the first to use a detonator (in Nobel's case a metal capsule filled with mercury fulminate) to initiate the decomposition of nitroglycerine. The reaction is:

$$C_3H_5N_3O_9 \rightarrow 3CO_2 + 2.5H_2O + 1.5N_2 + 0.25O_2 \tag{2.9}$$

This is accompanied by the release of 6700 J of heat per gramme of nitroglycerine liquid together with the evolution of 0.74 litres of gas (compared with only 0.265 litres for gunpowder). The explosive process is much more rapid than that which occurs in black powder because the reacting atoms are in much more intimate contact.

Nitroglycerine is an explosive of the nitrate ester type. Inspection of a more detailed representation of its molecular structure as shown below

$$
\begin{array}{l}
CH_2ONO_2 \\
| \\
CHONO_2 \\
| \\
CH_2ONO_2
\end{array}
\tag{2.10}
$$

indicates that it contains three of the sub-group $-O-NO_2$, characteristic of esters. Other important high explosive compounds containing this subgroup include nitroglycol, $C_2H_4O_6N_2$ (which is used as a substitute for nitroglycerine in the development of low-freezing point dynamites which remain plastic and stable at low temperatures), and pentaerythritol tetranitrate, commonly known as PETN of gross formula $C_5H_8O_{12}N_4$, which is used as a main charge in munitions and also in demolition detonators and detonating cord. The heats of explosion or mass specific energy Q of these materials are 6730 and 5940 kJ/kg respectively and a kilogramme of them generates 0.74 and 0.79 m^3 of gas respectively.

A second important category is the nitramine type of material. Here nitrogroups are linked to a nitrogen atom as illustrated below:

$$
\begin{array}{l}
-N- \\
| \\
NO_2
\end{array}
\tag{2.11}
$$

Important members of this class of explosive are:

1 1,3,5-trinitro-1,3,5-triaza-cyclohexane $(C_3H_6O_6N_6)$, more commonly known as cyclonite or RDX, which is used as a filling for military munitions and in special applications such as linear cutting charges;
2 1,3,5,7-tetranitro-1,3,5,7-tetra-aza-cyclo-octane $(C_4H_8O_8N_8)$, commonly known as HMX, which is a white, crystalline solid used as a filling for high performance munitions; and
3 N-methyl-N-nitro-2,4,6-trinitroaniline $(C_7H_5O_8N_5)$, commonly known as tetryl or CE which is used in military detonators.

Perhaps the most widely known explosive compound is TNT, a member of the nitro-explosive class which contains the $-NO_2$ group linked to a carbon atom, which was developed by the German Wilbrand in 1863. The full name is 2,4,6-trinitrotoluene of gross formula $C_7H_5O_6N_3$ which is made by the action of nitric and sulphuric acids on toluene (methyl benzene). It exists as a pale yellow-brown crystalline solid with a melting point of 80.6°C, a mass specific energy of 4520 kJ/kg and produces 0.73 m³ of gas per kilogramme. TNT is used as the filling for military munitions where it is usually mixed with other explosives such as RDX, HMX or ammonium nitrate. It also finds use as a component in pourable versions of some commercial blasting explosive mixtures such as ammonal (ammonium nitrate and aluminium powder) and slurry explosives (saturated aqueous solutions of ammonium nitrate and other nitrates).

As well as being a very effective explosive, TNT is widely accepted as the basis for comparison with other condensed high explosive materials as described in Chapter 3. It also offers a convenient way of quantifying the yield from gas and vapour cloud explosions as described in Chapter 4. The essence of the approach outlined in these chapters is that a particular explosion source is equated to a mass of TNT by considering the energy contained within the source. Thus, using an example from above, 1 kg of nitroglycerine containing 6700 kJ of energy is equivalent to ([6700/4520] × 1 =) 1.48 kg TNT.

(ii) Metal derivative compounds
When certain heavy metals such as lead or silver are treated with hydrazoic or styphnic acid, explosive compounds result. The most significant of these are lead azide (PbN_3), silver azide (AgN_3) and lead styphnate ($PbC_6H[NO_2]_3[OH]_2$). It should be noted that the azides contain no oxygen at all and, though the explosive performance of metal derivatives generally is inferior to CHON-type compounds, they do find an important role in initiation of CHON materials.

2.4 Oxygen balance

The stoichiometric oxidation described by Equation 2.1 is a special case of the decomposition of the explosive material. Generally, when an explosive material reacts, the first stage in decomposition can be written

$$C_aH_bN_cO_d \rightarrow aC + bH + cO + dN \qquad (2.12)$$

Which gases will form as a consequence of decomposition will depend on the number of atoms of oxygen (d in Equation 2.12) compared with the number of carbon and hydrogen atoms ($a + b$). If d is large compared to ($a + b$) then the products will be carbon dioxide (CO_2), water (H_2O), carbon monoxide (CO) and nitrogen (N). If d is somewhat smaller relative to ($a + b$) products will be carbon monoxide, water and nitrogen. If there is only a little oxygen available in the explosive, products will be carbon monoxide, hydrogen, nitrogen and free carbon appearing as smoke.

The so-called 'oxygen balance' can be assessed by considering the number of molecules of carbon and hydrogen relative to the number of molecules of oxygen present. Two oxygen atoms will fully oxidise one carbon atom while half an oxygen atom is needed to oxidise a hydrogen atom. The oxygen balance, usually written Ω, is proportional to the difference between the number of oxygen molecules present in the compound d, and the quantity $(2a + \frac{1}{2}b)$ representing the number of atoms with which oxygen will combine in a stoichiometric reaction. This leads to the definition of oxygen balance as

$$\Omega = \frac{(d - 2a - b/2) \times 16}{M} \times 100\% \tag{2.13}$$

where M is the molecular weight of the explosive compound and 16 is the atomic weight of oxygen. Therefore, in a stoichiometric reaction, Ω will be zero and the production of heat energy per unit mass of material will be maximised. This situation is almost realised in nitroglycerine with gross formula $C_3H_5N_3O_9$ where d is 9, a is 3 and b is 5. The molecular weight of nitroglycerine is 227, so Ω is + 3.5% indicating that a near balance has been achieved showing that nitroglycerine is an effective explosive.

On the other hand many military explosives are oxygen deficient, showing a large negative oxygen balance. For instance, TNT with gross formula $C_7H_5N_3O_6$, and which also has a molecular weight of 227, has an oxygen balance of − 74%. With such a large deficit, it is common in military munitions to add a material with a significant positive oxygen balance such as ammonium nitrate (Ω = + 20%) to redress the situation and enhance the reaction.

Table 2.1 shows the oxygen balance of a number of common explosives and explosive components.

Table 2.1

Material	Oxygen balance $\Omega\%$
Ammonium nitrate NH_4NO_3	+20.0
Potassium nitrate KNO_3	+39.6
Nitroglycerine $C_3H_5N_3O_3$	+3.5
Nitrocellulose (gun-cotton) $C_{12}H_{14}N_6O_{22}$	−28.6
Tetryl $C_7H_5N_5O_8$	−47.4
TNT $C_7H_5N_3O_6$	−74.0
Dinitrotoluene $C_7H_6N_2O_4$	−114.4
Nitroguanidine $CH_4N_4O_2$	−38.7
RDX (cyclonite) $C_3H_6N_6O_6$	−21.6

2.5   Explosion processes

2.5.1   Terminology

Combustion is the term used to describe any oxidation reaction, including those requiring the presence of oxygen from outside as well as those that use oxygen which is an integral part of the reacting compound. When applied to the particular case of the oxidation of gun propellants, the process is usually denoted as *burning*.

In the case of explosive materials, they can decompose at a rate much below the speed of sound in the material and the combustion process is known as *deflagration*. Deflagration is propagated by the liberated heat of reaction: the flow direction of the reaction products is in opposition to the direction of decomposition.

Detonation is the form of reaction of an explosive which produces a high intensity shock wave. Most explosives can be detonated if given sufficient stimulus. The reaction is accompanied by large pressure and temperature gradients at the shock wavefront and the reaction is initiated instantaneously. The reaction rate, described by the detonation velocity, lies between about 1500 and 9000 m/s which is appreciably faster than propagation by thermal conduction and radiation as in the deflagration process described above.

2.5.2 Transition to detonation

When an explosive is undergoing decomposition by burning, the reaction is proceeding at or just above the surface of the solid material layer by layer as each is brought to the ignition temperature of the material. It does this by the transfer of heat into the solid material from the reaction zone. The rate at which heat is transferred into the solid influences the rate at which the surface of the solid recedes. The main influence on the linear burning rate r is the surface pressure P, with a large surface pressure leading to a narrower reaction zone and hence a faster transfer of heat to the solid material. The relationship between r and P may be written in simplified form as

$$r = \beta P^{\alpha} \tag{2.14}$$

where α and β are empirical constants particular to a material. For the case of a typical gun propellant which will burn at about 5 mm/s under normal atmospheric conditions, when in a gun barrel (where the pressure may be up to 4000 atmospheres) this rate can increase to about 400 mm/s. If this rate can be further increased, detonation may result. This is unlikely to occur for propellants, however, because since α is less than 1 the burning rate-pressure curve is of the form shown in Figure 2.1(a) with the slope of the curve reducing as pressure increases. Other explosive materials, on the other hand, have α greater than unity, in which case the curve of Figure 2.1(b) results and burning rate accelerates dramatically with increased confining pressure.

The result of the accelerating flame front could be a burning velocity in excess of the sound speed in the material and the development of detonation.

An alternative method of provoking detonation is to impact the explosive material with a high velocity shock wave derived from a mechanical or (most likely) another explosive source such as a primary explosive which can be detonated by a relatively weak mechanical shock or a spark. The shock wave hits the explosive to be detonated, compresses it and causes it to undergo adiabatic heating. This process liberates energy as the explosive exothermically decomposes causing shock wave acceleration.

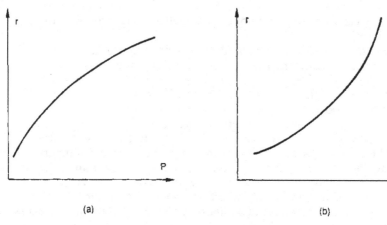

Figure 2.1   *Effect on reaction rate of confining pressure*

When the velocity of the wave exceeds the speed of sound in the explosive medium, detonation results at a velocity that is a characteristic of the explosive and is independent of pressure (unlike in the case of burning described above). Generally, in practical charges, although this process requires both a finite time and finite travel distance to develop, any delay is of little practical significance. It should be noted, however, that below a critical charge size, detonation may not be achievable.

2.5.3   Detonation

Steady detonation may be considered as a self-propagating process where the effect of compression of the shock front discontinuity changes the state of the explosive so that the exothermic reaction is established with the requisite velocity. The increases in pressure, temperature and density can occur during a time interval as short as a few picoseconds in a homogeneous liquid explosive in a reaction zone with a thickness of about 200 microns. Within this zone the pressure can be in excess of 200 kilobars, the temperature above about 3000 K and density can be about 1.3 times the initial value.

The analytical process is described here in outline for the idealised situation of plane stationary detonation in which an explosive particle passes through a pressure process which is the same for all particles involved in the detonation event. At the rear of the reaction zone, pressure is maintained by the acceleration of the reaction products as they expand backwards relative to the reaction front. Here, because most chemical energy has been liberated, flow is unsteady. The propagation velocity of pressure disturbances at the rear of the reaction zone is less than the difference between the detonation velocity and the local particle velocity and, as a consequence, the detonation front moves quickly away from the rear flow area developing a pressure profile which becomes successively less steep.

At the front of the reaction zone, however, the velocity of expansion of pressure disturbances exceeds the difference between the detonation velocity and local particle velocity. This means that pressure contributions from the chemical reaction energy continually reinforce the shock front. To maintain a steady flow, the pressure profile changes to accommodate the energy liberated at every point in the reaction zone.

The boundary between the steady and unsteady flow regions is the plane in which the velocity with which pressure disturbances expand (the sound velocity in the explosively produced gases), a_g, equals the difference between detonation velocity, D, and the local particle velocity just behind the detonation wavefront u_d. This is called the Chapman-Jouguet (or CJ) plane.[1,2] Thus, we can write

$$D - u_d = a_g \tag{2.15}$$

The definition of an ideal detonation is a plane detonation wave in which reaction is completed in the CJ plane. To illustrate these ideas more analytically using the approach of Reference 4, consider a plane shock wave as shown in Figure 2.2.

If a control volume moving with the detonation wavefront at velocity D is established as shown, from the law of mass conservation we can write, for unit area of wavefront

$$\rho_o D = \rho_1 (D - u_d) \tag{2.16}$$

From conservation of momentum we have

$$p_1 - p_o = \rho_o D u_d \tag{2.17}$$

and from the law of conservation of energy we can write

$$\frac{p_1}{\rho_1} + \frac{(u_d - D)^2}{2} + e_1 = \frac{p_o}{\rho_o} + \frac{D^2}{2} + e_o \tag{2.18}$$

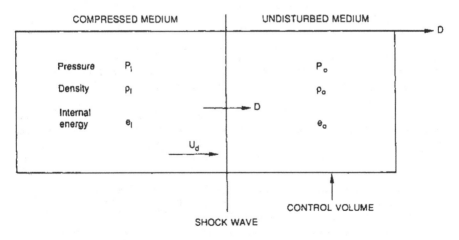

Figure 2.2 *Pressure and other changes across a plane shock wave*

In these equations, subscript o refers to the undisturbed medium ahead of the detonation wave and subscript 1 refers to the medium in shock compression. Equation 2.18 can be expanded and simplified to yield

$$e_1 - e_o + \frac{u_d^2}{2} = \frac{p_o}{\rho_o} - \frac{p_1}{\rho_1} + u_d D \qquad (2.19)$$

From Equation 2.16, the density behind the detonation wave can be written

$$\rho_1 = \frac{\rho_o D}{D - u_d} \qquad (2.20)$$

and from Equation 2.17, p_o can be written as

$$p_o = p_1 - \rho_o D u_d \qquad (2.21)$$

If Equations 2.20 and 2.21 are substituted into Equation 2.19, after simplification the resulting equation is

$$p_1 u_d = \rho_o D \left(e_1 - e_o + \frac{u_d^2}{2} \right) \qquad (2.22)$$

By writing specific volume v as $1/\rho$, Equations 2.16, 2.17 and 2.22 can be combined. The result is the Hugoniot equation

$$e_1 - e_o = \tfrac{1}{2}(p_1 + p_o)(v_o - v_1) \qquad (2.23)$$

This can be modified to include the energy of an explosive reaction Q to give

$$e_1 - e_o = \tfrac{1}{2}(p_1 + p_o)(v_o - v_1) + Q \qquad (2.24)$$

Equation 2.23 can be plotted as shown in Figure 2.3 where $p_1 = p$ and $v_1 = v$. Equations 2.16 and 2.17 can be recast to give

$$p_1 - p_o = (v_o - v_1)\rho_o^2 D^2 \qquad (2.25)$$

which represents a straight line on Figure 2.3 of slope $(-\rho_o^2 D^2)$, called the Rayleigh line, which is tangential to the Hugoniot line at the Chapman-Jouguet (CJ) point. The portion of the Hugoniot curve to the left of the line $v = v_o$ represents the region of stable detonation, while the part of the curve below the line $p = p_o$ represents deflagration. Between these two limits neither detonation nor deflagration is theoretically possible. Physically this means that there is no continuous transition between deflagration and detonation. In practice, however, this transition does occur in both directions.

Since the medium here is explosive, the explosive reaction is, of course, coincident with the shock wavefront and the propagation of the wave will be maintained by the energy released by the reaction. Equations 2.17 to 2.23 can be simplified to account for the physical situation. Pressure p_1 may be called the detonation pressure. Since ρ_1 is the density of the explosive gases behind

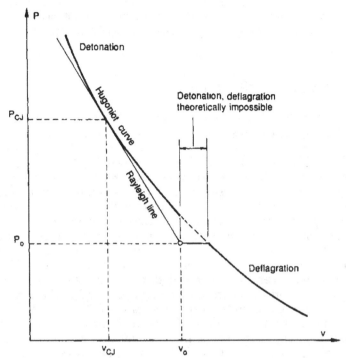

Figure 2.3   *Detonation and deflagration: Hugoniot equation and Rayleigh line (after Ref. 4)*

the shock front which is much higher than the density of the explosive ρ_0, Equation 2.17 can be simplified to give detonation pressure as

$$p_1 = \rho_0 D u_{\mathrm{d}} \tag{2.26}$$

This means that detonation pressure increases with explosive density. This is an important consideration if the explosive is to be used as an in-contact charge to make use of the 'brisance' or 'hammer-blow' effect of the detonation wave (see below). Conversely, for blasting of soft rock or generating a milder thrust effect, a loosely textured and therefore less dense explosive would be appropriate.

2.5.4   Velocity of detonation

The denser the explosive material, the faster will be the speed of the detonation wave. The greatest velocity is associated with charges comprising a single crystal of explosive since such material will have the highest possible density. In practice, charge densities are lower than the maximum and, provided the charge is not too small and is well confined, the detonation velocity D_{I} (m/s) of a charge of density ρ_{I} (kg/m^3) is related to the detonation velocity D_{II} of a charge of density ρ_{II} by the empirical equation

$$D_{\mathrm{I}} = D_{\mathrm{II}} + 3.5 \times 10^6 (\rho_{\mathrm{I}} - \rho_{\mathrm{II}}) \tag{2.27}$$

It is worth noting that there is a lower limit to the diameter of a charge, the so-called critical diameter, below which detonation cannot become established. For materials in military explosives such as RDX, this diameter is approximately 10 mm.

2.6   Explosives classification

As has been noted above, explosives can either detonate or deflagrate. High explosives detonate to create shock waves, burst or shatter materials in or on which they are located, penetrate materials, produce lift and heave of materials and, when detonated in air or under water, produce airblast or underwater pressure pulses. Low explosives like propellants burn to propel projectiles such as artillery shells, small-arms rounds and rockets. Pyrotechnics are low explosives which burn to ignite propellants, produce delays in an explosive train or create heat, smoke, fire and noise.

Generally, classification of these materials is on the basis of their sensitivity to initiation. A primary explosive is one that can be easily detonated by simple ignition from a spark, flame or impact. Materials such as mercury fulminate, lead and silver azide and lead styphnate are all primary explosives. These are the sort of materials that might be found in the percussion cap of gun ammunition. Secondary explosives can be detonated, though less easily than primary explosives. Examples of these materials include TNT, RDX and tetryl among many others. In the example of gun ammunition, a secondary explosive would be used for the main explosive charge of the shell.

In order to achieve the required properties of safety, reliability and performance, it is common practice in both military and commercial explosives manufacture to blend explosive compounds. Economic considerations are, of course, also important in any formulation. For commercial use, explosives are generally made from the cheaper ingredients: TNT or nitroglycerine might be mixed with low-cost nitrates, for example. One penalty incurred by this approach is that such material has a generally short shelf-life. At the other end of the spectrum, military explosives are composed of more expensive ingredients such as binary mixtures of stable compounds like TNT and RDX or HMX with TNT. The requirement here is for a long shelf-life coupled with safety in handling and reliability of performance.

2.7   Initiation

All explosive materials are sensitive to heat and various methods used to initiate explosives make use of heat energy. The reliability of an explosive depends on its thermal instability, while conversely its safety depends on its thermal stability. It is clear that, for development of a successful explosive material, a balance between these two conflicting aspects must be struck.

As a charge is heated its temperature rises. If the rate at which heat is supplied is greater than the rate at which heat is dissipated, a runaway reaction could be developed leading to ignition of the entire charge. If the temperature of the charge rises to between about 150 and 350°C, common explosive materials are likely to ignite. The basis of all systems which are used to initiate explosives deliberately is the supply of heat energy to the material. If the activation energy, the quantity E in the diagrammatic representation of an explosive reaction shown in Figure 2.4, is provided, the reaction starts and decomposition to yield the heat of explosion Q will occur.

There are a number of ways that the necessary heat can be supplied. External heating could be used by holding the explosive in a container and warming it: this method is not of practical significance, being slow and unpredictable in outcome. It is worth noting that direct application of a flame is essential for propellant and pyrotechnics initiation because they function through burning. However, for secondary explosives this method is unreliable because initial deflagration may not undergo transition to detonation.

Most primary explosives can be initiated by percussion. Two processes are active in bringing the explosive to ignition temperature: air cavities within the explosive are compressed producing a temperature rise and frictional heating between explosive grains. The degree of sensitivity required of a primary explosive is fairly critical if reliability and safety levels are to be acceptable. It should be emphasised that percussive initiation is an ever-present hazard in explosive manufacture and handling. As noted above, frictional heating can cause initiation as part of the percussion process. Practical methods based on deliberate generation of frictional heat are few: the safety match and the 'snap' in a Christmas cracker are examples. However, if percussion and friction are combined by rapidly pushing a

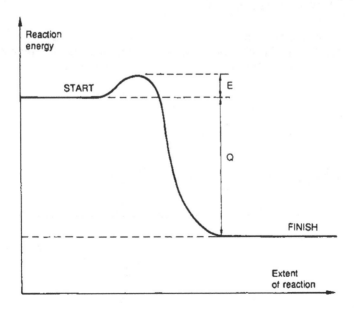

Figure 2.4 *Energy changes in explosion reaction*

small needle into a primary explosive, this method of initiation is known as 'stabbing' which gives very reliable results.

There are several ways of converting electrical energy to heat energy based on resistance heating. The most common is the bridgewire in which a filament is heated to ignite an attached small quantity of explosive. In the exploding bridgewire method a high tension pulse causes a filament to vaporise producing both high temperature and a shock wave which is often sufficient to initiate a secondary explosive. Finally, in a so-called 'conducting cap', a primary explosive is mixed with carbon to make it conducting but of high resistance: a low tension source produces an essentially instantaneous temperature rise to ignition.

As noted above, initiation to detonation of a secondary explosive is almost always by exposing it to a high intensity shock wave generated by the detonation of another explosive material, ideally in contact with the (larger) charge to be initiated. In the electric demolition detonator a number of the initiation methods described above are used in conjunction to produce the necessary pulse. Such as device is shown in Figure 2.5.

In this device (which finds application in both military and commercial fields) the electrically energised bridgewire ignites the 'match-head'-type compound adhering to it. This, in turn, ignites the primary explosive which burns and undergoes transition to detonation. The fairly sensitive secondary explosive in the detonator is now initiated and produces a shock of sufficient intensity to initiate the main charge (a secondary explosive of somewhat lower sensitivity) in contact with the detonator.

2.8   Effects of explosives

If a detonating explosive is in contact with a solid material the arrival of the detonation wave at the surface of the explosive will generate intense stress waves in the material, producing crushing and shattering disintegration of

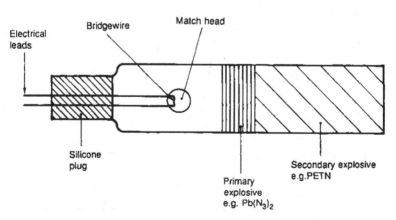

Figure 2.5   *Electric detonator*

the material. This hammer-blow effect is called 'brisance'. The dynamic pressure at the detonation wavefront is the detonation pressure p_1 given in kilobars by the empirical equation

$$p_1 = 2.5 \times 10^{-9} \rho D^2 \tag{2.28}$$

where ρ (kg/m^3) is explosive density and D (m/s) is detonation wave speed. Thus if D is 7400 m/s for RDX which is of density 1500 kg/m^3 then the detonation pressure p_1 is, by Equation 2.28, 205.4 kilobars, a pressure far in excess of the compressive strength of most materials.

If the explosive is surrounded by a medium such as air a blast wave will be created (see Chapter 3). In water not only is a shock wave transmitted but a rapidly expanding gas bubble is formed (see Chapter 5). In earth, ground shock waves are generated and a crater may be formed (see Chapter 7). These effects are derived from the work capacity of the explosive usually referred to as the 'power' of the explosive. Power is the useful chemical energy of the explosive often referred to as its strength. It is proportional to the number of moles of gas produced and the heat of explosion Q. The heat of explosion and the gas volume per gramme of explosive (V) for a number of primary and secondary explosives are given in Table 2.2.

Table 2.2

	Q (kJ/kg)	V (cm^3/g)	PI
Primary explosives			
Lead styphnate (C$_6$H$_3$N$_3$O$_9$Pb)	1885	325	21
Lead azide (Pb[N$_3$]$_2$)	1610	230	13
Secondary explosives			
Nitroglycerine	6275	740	159
RDX	5130	908	159
TNT	4080	790	117
RDX/TNT (60/40)	4500	796	138
HMX	5130	910	160

Explosive power is generally calculated as the product of Q and V and compared with the product of Q and V for picric acid (3745 kJ/kg and 790 cm^3/g respectively) which is 2.959 × 10^6 to form a power index PI defined as

$$PI = \frac{(QV)_{\text{explosive}}}{(QV)_{\text{picric}}} \tag{2.29}$$

As Table 2.2 shows, the power index for secondary explosives is much greater than for primary explosives. This means, of course, that secondary explosives will be much more effective in producing lifting and heaving of soils and rocks, generating high intensity airblast waves and in creating intense underwater shock and bubble pulses.

A special use of high explosives is in hollow or shaped charges in which the geometry of the detonation wave is altered to produce greater directionality in the way that explosive energy is delivered. The effect was first noted by the American chemist Munroe who discovered that by hollowing out the end of a dynamite stick which was then placed in contact with a solid material, the crater produced was far deeper than if the end of the dynamite stick remained flat. This 'Munroe Effect' has been the subject of extensive research in the past century and is the principle behind a whole range of military and civilian explosive devices used for attacking armour or for cutting materials in demolition applications. Discussion of the design and performance of these 'shaped charges' is beyond the scope of this chapter and reference should be made to Chapter 12 or to one of a number of other sources available.[3]

Given below are a number of references in the general field of explosives which should be consulted for a more detailed treatment.[4,5,6] In particular, Reference 7 presents a concise overview of a wide field.

2.9 References

[1] Chapman D.L., On the rate of explosion in gases. *London, Dublin and Edinburgh Philosophical Magazine.* **213**, Series 5 (47), 90. (1899)
[2] Jouguet E., Sur la propagation des reactions chimiques dans la gaze. *Journal of Pure and Applied Mathematics* **70**, Series 6 (1), 347. (1905)
[3] Walters W.P., Zukas J.A., *Fundamentals of shaped charges.* Wiley, New York (1989)
[4] Meyer R., *Explosives.* (3rd Edition) VCH (1987)
[5] Johansson C.H., Persson P.A., *Detonics of High Explosives.* Academic Press, London (1970)
[6] Kaye S.H., (Ed) *Encyclopaedia of Explosions and Related Terms.* Vols 1–12, US Research and Development Command Large Calibre Weapons Systems Laboratory, New Jersey (1983)
[7] Bailey A. Murray S.G., *Explosives, Propellants and Pyrotechnics.* Brassey's, London (1989)

List of symbols

a_g	velocity of sound in products of reaction
D	detonation velocity
D_I	detonation velocity in material 1
D_{II}	detonation velocity in material 2
e_o	specific internal energy of undisturbed medium
e_1	specific internal energy of compressed medium
ΔG	Gibbs free energy change
ΔH	heat of explosion
P	surface pressure of burning explosive
p_o	pressure in undisturbed medium
p_1	pressure behind wavefront in compressed medium
PI	power index
Q	heat of explosion

r	linear burning rate
ΔS	entropy change
T	temperature
u_d	particle velocity behind wavefront
V	gas volume (in cc) produced per gramme of explosive
v_o	specific volume of undisturbed medium
v_1	specific volume of compressed medium
α	empirical burning rate constant
β	empirical burning rate constant
ρ_I	density of explosive material 1
ρ_{II}	density of explosive material 2
ρ_o	density of undisturbed medium
ρ_1	density of compressed medium
Ω	oxygen balance

3 Blast waves and blast loading

3.1 Introduction

As has been described in Chapter 2, chemical or nuclear explosions liberate large amounts of energy which, in heating the surrounding medium, produces very high local pressures. This pressure disturbance, moving outwards, 'shocks up' and develops into a blast wave. When a condensed high explosive material detonates almost 100% of the energy liberated is converted into blast energy. While the nature of blast is similar in a nuclear explosion only about 50% of the energy produces blast; the remainder is in the form of thermal and other types of radiation.

In the following sections blast waves produced by the detonation of high explosive materials and nuclear devices in free air and on the ground are discussed with particular reference to conditions at the blast wavefront.

3.2 Blast waves in air from condensed high explosives

When a condensed high explosive is initiated the following sequence of events occurs. Firstly, the explosion reaction generates hot gases which can be at pressures from 100 kilobars up to 300 kilobars and at temperatures of about 3000–4000°C. A violent expansion of these explosive gases then occurs and the surrounding air is forced out of the volume it occupies. As a consequence a layer of compressed air – the blast wave – forms in front of these gases containing most of energy released by the explosion. As the explosively formed gases expand their pressure falls to atmospheric pressure as the blast wave moves outwards from the source. The pressure of the compressed air at the blast wavefront also falls with increasing distance. Eventually, as the explosive gases continue to expand they cool and their pressure falls a little below atmospheric pressure. This is because even though the static pressure of these gases may be atmospheric, since the gas molecules have mass and are moving it takes a little longer and a further distance of travel before their momentum is destroyed. The gases are now 'over-expanded' and the result is a reversal of flow towards the source driven by the (small) pressure differential between atmospheric conditions and the pressure of the gases. The effect on the blast wave shape is to induce a region of 'underpressure' where pressure is below atmospheric: this is the so-called negative phase. Eventually the situation returns to equilibrium as the air and gases pushed away from the source return.

The equations of motion that are needed to describe a blast wave are complex and were first solved numerically by Brode[1] and subsequently verified experimentally by Kingery[2]. An outline of the analysis is provided below which is a development of the ideas presented in Chapter 2. The reader is recommended to refer to a specialist compressible gas dynamics text for a more comprehensive treatment.

3.3 Blast wave equations for spherical charges of condensed high explosive.

In order to make the problem more approachable, the following discussion will treat the formation, description and effects of blast waves in three locations. The first location is within the explosive material itself where the detonation of the explosive material occurs. The second is the region outside the explosive in the surrounding medium where the blast wave is formed. Finally, in subsequent chapters, the region of interaction will be considered where the blast wave generates a structural loading.

3.3.1 Inside the charge

The following analysis is adapted from that presented by Henrych[3] and complements the analysis of a plane detonation wave presented in Chapter 2.

Consider a spherical charge of condensed high explosive of radius R_c as shown in Figure 3.1. The charge has been centrally initiated and the detonation wavefront is at radius R_d. The reaction zone is a thin region behind this wavefront.

With reference to Figure 3.1, in zone A (where reaction is complete), pressure p, density ρ, particle velocity u and temperature T are all varying with radius from the charge centre r and with time t. The exact relationships between, say, pressure, radius and time depend on the type of explosive comprising the charge. In zone B ahead of the thin reaction zone and thus just beyond the detonation wavefront of radius R_d, the unreacted explosive is at ambient pressure and temperature p_{ex} and T_{ex} respectively and is at rest with zero particle velocity. Its internal energy per unit mass (the specific energy) is e_{ex} and the velocity of sound in the unreacted explosive is a_{ex}. Zone C is at the surrounding atmospheric conditions so pressure is atmospheric p_o, temperature is T_o, particle velocity is zero, the air has specific energy e_o and sound velocity is a_o.

To obtain the equations necessary to evaluate pressure and particle velocity variations etc., consider an elemental control volume at radius r within the exploded or reacted material as shown in Figure 3.2.

The requirement for conservation of mass means that, for the control volume, the mass entering less mass leaving must equal any mass retained due to density change. Therefore we can write

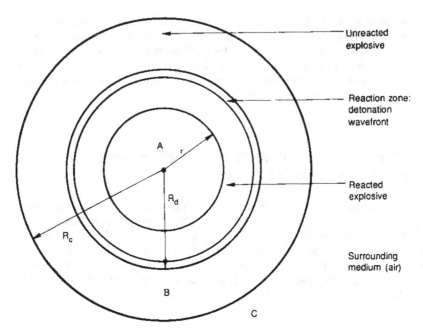

Figure 3.1   *Inside the charge (after Ref. 3)*

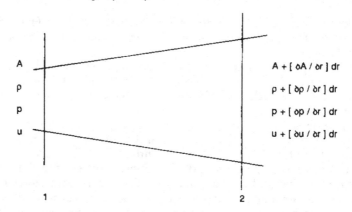

Figure 3.2   *Elemental control volume (after Ref. 3)*

$$Au\rho \, \partial t - \left[A + \frac{\partial A}{\partial r} dr\right]\left[u + \frac{\partial u}{\partial r} dr\right]\left[\rho + \frac{\partial \rho}{\partial r} dr\right] = A\partial r \frac{\partial \rho}{\partial t} dt \tag{3.1}$$

where A is the area of section 1 of the control volume. This yields on simplification

$$\frac{\partial \rho}{\partial t} + u\frac{\partial \rho}{\partial r} + \rho\frac{\partial u}{\partial r} + \frac{\rho u}{A}\frac{\partial A}{\partial r} = 0 \tag{3.2}$$

The requirement for conservation of momentum means that the sum of all the forces acting on the material in the control volume must equal the product of mass and acceleration for the element

$$pA - \left[p + \frac{\partial p}{\partial r}dr\right]\left[A + \frac{\partial A}{\partial r}dr\right] + \left[p + \frac{\partial p}{\partial r}\frac{dr}{2}\right]dA = A\rho \, dr\frac{du}{dt} \tag{3.3}$$

where the first term is the force on face 1 in Figure 3.2, the second term is the force on face 2 and the third term is the force on the surface between 1 and 2 resolved in the radial direction. Equation 3.3, on simplifying yields,

$$A\frac{\partial p}{\partial r}dr = A\rho \, dr\frac{du}{dt} \tag{3.4}$$

Then noting that we can write

$$du = \left[\frac{\partial u}{\partial r}\right]dr + \left[\frac{\partial u}{\partial t}\right]dt \tag{3.5}$$

and that $u = dr/dt$ so that

$$\frac{du}{dt} = u\frac{\partial u}{\partial r} + \frac{\partial u}{\partial t} \tag{3.6}$$

by substitution into Equation 3.4 we obtain

$$\frac{\partial u}{\partial t} + u\frac{\partial u}{\partial r} + \frac{1}{\rho}\frac{\partial p}{\partial r} = 0 \tag{3.7}$$

By the principle of conservation of energy which states that for unit mass of material, an increment of heat dQ equals an increment of internal energy de plus any work done in changing volume of the material pdv where v is specific volume, the first law of thermodynamics states

$$dQ = de + pdv \tag{3.8}$$

Since specific volume is

$$v = \frac{1}{\rho} \quad \text{then} \quad \frac{dv}{d\rho} = -\frac{1}{\rho^2}$$

$$\therefore \quad dv = -\frac{d\rho}{\rho^2} \tag{3.9}$$

then Equation 3.8 becomes

$$dQ = de - \frac{p}{\rho^2}d\rho \tag{3.10}$$

and, by dividing through by T we obtain

$$\frac{dQ}{T} = \frac{de}{T} - \frac{p}{\rho^2}\frac{d\rho}{T} = dS \tag{3.11}$$

where dS is the change in entropy of the system. Then by dividing by dt and rearranging we have

$$T\frac{dS}{dt} = \frac{de}{dt} - \frac{p}{\rho^2}\frac{d\rho}{dt} \tag{3.12}$$

We also need an equation of state which may be written (though other forms are possible) as

$$\frac{p}{\rho} = RT \tag{3.13}$$

where R is the gas constant, together with an equation for the internal energy of the system

$$e = C_v T - \int_{\rho_c}^{\rho}\left[T\left[\frac{\partial p}{\partial T}\right]_\rho - p\right]\frac{d\rho}{\rho^2} \tag{3.14}$$

where C_v is the specific heat at constant volume. We now have a set of five equations but with six unknowns: pressure, density, particle velocity, temperature, internal energy and entropy. So, if solution is to proceed, a sixth equation is required and some assumption must be made about the sixth relationship which is of the general form

$$f(\rho, p, u, T, r, t) = 0 \tag{3.15}$$

A particular form which relates to the detonation process we are describing could be that of Chapman-Jouguet (Equation 2.15)

$$D - u_d = a_g \tag{3.16}$$

where D is the detonation wave speed (approximately 7000 m/s in TNT), u_d is particle velocity behind the detonation wavefront and a_g is the local sound velocity in the explosive gases.

Boundary conditions for the solution of the six equations are known: at all time particle velocity at $r = 0$ is zero, the pressure at the detonation wavefront is p_1 where density is ρ_1 and particle velocity is u_d. Thus

$$u(0, t) = 0 \quad p(R_d, t) = p_1$$

$$\rho(R_d, t) = \rho_1 \quad u(R_d, t) = u_d \tag{3.17}$$

Therefore it is possible for numerical solution to proceed to yield results for the six parameters.

3.3.2 Outside the charge

The system of equations above refers to the detonation process. In Chapter 2 a similar approach was adopted for a plane wavefront. Blast wave parameters can be evaluated in an exactly similar manner. The situation now is as shown in Figure 3.3.

In zone I, the region of the expanding explosive gases, motion can be expressed by the analysis of Equations 3.1 to 3.15 above. In zone II, the region of the surrounding compressed air, motion is again similar so these equations can be used, though it should be noted that the equation of state depends on the medium being considered. Also, Equation 3.15 may be

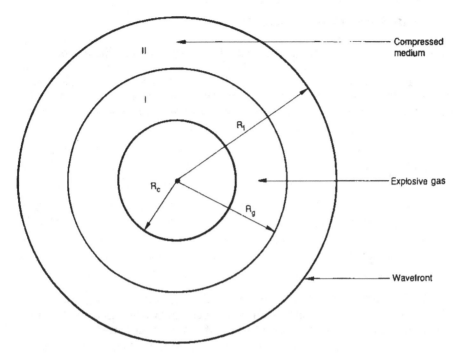

Figure 3.3 *Outside the charge (after Ref. 3)*

different. As before, boundary conditions can be precisely defined. Particle velocity at $r = 0$ is always zero. The interface between the explosively produced gases and the tail of the blast wave (at $r = R_g$) has the same particle velocity equal to the rate at which the interface expands. The pressure at the interface is the same in both media though the densities are not. At the blast wavefront at radius R_f, pressure is p_s and the corresponding particle velocity is u_s. Thus we have

$$u(0, t) = 0 \quad p_g(R_g, t) = p_m(R_g, t)$$

$$u_g(R_g, t) = u_m(R_g, t) = \frac{dR_g}{dt}$$

$$\rho_g(R_g, t) \neq \rho_m(R_g, t) \quad p(R_f, t) = p_s$$

$$u(R_f, t) = u_s \tag{3.18}$$

where suffix g refers to the explosive gases, suffix m refers to the surrounding medium and suffix s refers to the blast wavefront. Thus, once again we have obtained six equations for six unknowns which can be solved numerically.

3.4   Blast wavefront parameters

Of particular importance are the wavefront parameters. These were first presented by Rankine and Hugoniot in 1870[4] to describe normal shocks in ideal gases and are available in a number of references such as Liepmann and Roshko.[5] The equations for blast wavefront velocity U_s, air density behind the wavefront ρ_s, and the maximum dynamic pressure q_s are given below

$$U_s = \sqrt{\frac{6p_s + 7p_0}{7p_0}} \cdot a_0 \tag{3.19}$$

$$\rho_s = \frac{6p_s + 7p_0}{p_s + 7p_0} \cdot \rho_0 \tag{3.20}$$

$$q_s = \frac{5p_s^2}{2(p_s + 7p_0)} \tag{3.21}$$

where p_s is peak static overpressure, p_0 is ambient air pressure ahead of the blast wave, ρ_0 is the density of air at ambient pressure ahead of the blast wave and a_0 is the speed of sound in air at ambient pressure.

From Equation 3.21 taking p_0 as 101 kPa gives the results of Table 3.1 showing the variation of static pressure and dynamic pressure in a blast wave in free air.

Table 3.1

p_s (kPa)	q_s (kPa)
200	110
350	290
500	518
650	778

It will be seen that up to a peak static overpressure of about 500 kPa the value of dynamic pressure is numerically less than p_s. At overpressures in excess of about 500 kPa dynamic pressure exceeds static pressure. Equations 3.19 and 3.21 are plotted in Figures 3.4 and 3.5.

Brode's analysis leads to the following results for peak static overpressure p_s in the near field (when p_s is greater that 10 bar) and in the medium to far field (p_s between 0.1 and 10 bar).

$$p_s = \frac{6.7}{Z^3} + 1 \text{ bar} \quad (p_s > 10 \text{ bar})$$

$$p_s = \frac{0.975}{Z} + \frac{1.455}{Z^2} + \frac{5.85}{Z^3} - 0.019 \text{ bar} \quad (0.1 < p_s < 10 \text{ bar}) \tag{3.22}$$

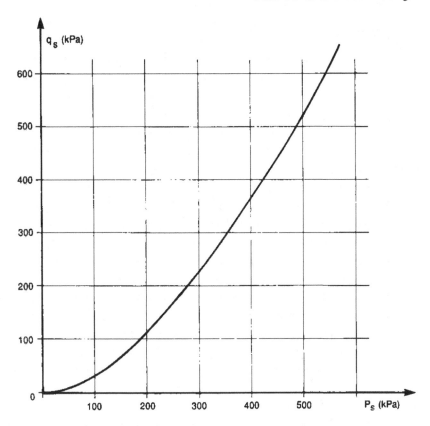

Figure 3.4 *Dynamic pressure vs side-on overpressure*

Here Z is scaled distance given by

$$Z = \frac{R}{W^{1/3}} \qquad\qquad (3.23)$$

where R is the distance from the charge centre in metres and W is the charge mass expressed in kilogrammes of TNT. The use of TNT as the 'reference' explosive in forming Z is universal. The first stage in quantifying blast waves from sources other than TNT is to convert the actual mass of the charge into a TNT equivalent mass. The simplest way of achieving this is to multiply the mass of explosive by a conversion factor based on its specific energy and that of TNT. Conversion factors for a range of explosives are shown in Table 3.2, adapted from Baker et al.[6]

Thus, a 100 kg charge of RDX converts to 118.5 kg of TNT since the ratio of the specific energies is 5360/4520 (= 1.185).

An alternative approach makes use of two conversion factors where the choice of which to use depends on whether the peak overpressure or the impulse delivered is to be matched for the actual explosive and the TNT equivalent. Table 3.3, adapted from the document TM5-855-1, gives examples of these alternative conversion factors.

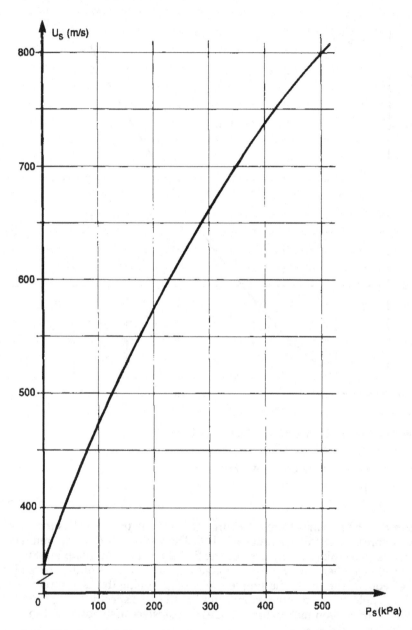

Figure 3.5 *Wavefront velocity vs side-on overpressure*

Other solutions exist for wavefront parameters from both numerical solution and from experimental measurements. The equations below are given by Henrych[3] and are of a similar form to those of Brode.

Table 3.2

Explosive	Mass Specific energy Q_x(kJ/kg)	TNT Equivalent (Q_x/Q_{TNT})
Amatol 80/20 (80% ammonium nitrate 20% TNT)	2650	0.586
Compound B (60% RDX, 40% TNT)	5190	1.148
RDX (Cyclonite)	5360	1.185
HMX	5680	1.256
Lead azide	1540	0.340
Mercury fulminate	1790	0.395
Nitroglycerin (liquid)	6700	1.481
PETN	5800	1.282
Pentolite 50/50 (50% PETN 50% TNT)	5110	1.129
Tetryl	4520	1.000
TNT	4520	1.000
Torpex (42% RDX, 40% TNT, 18% Aluminium)	7540	1.667
Blasting gelatin (91% nitroglycerin, 7.9% nitrocellulose, 0.9% antacid, 0.2% water)	4520	1.000
60% Nitroglycerin dynamite	2710	0.600

Table 3.3

Explosive	Equivalent pressure factor	Equivalent impulse factor
Composition B (60% RDX, 40% TNT)	1.11	0.98
PETN	1.27	Not available
Pentolite	1.40	1.07
Tetryl	1.07	Not available
TNT	1.00	1.00

$$p_s = \frac{14.072}{Z} + \frac{5.540}{Z^2} - \frac{0.357}{Z^3} + \frac{0.00625}{Z^4} \text{ bar} \quad (0.05 \leq Z < 0.3)$$

$$p_s = \frac{6.194}{Z} - \frac{0.326}{Z^2} + \frac{2.132}{Z^3} \text{ bar} \quad (0.3 \leq Z \leq 1)$$

$$p_s = \frac{0.662}{Z} + \frac{4.05}{Z^2} + \frac{3.288}{Z^3} \text{ bar} \quad 1 \leq Z \leq 10 \quad (3.24)$$

The accuracy of predictions and measurements in the near field is somewhat lower than in the medium to far field probably because of the complexity of the flow processes involved in forming the blast wave close to the charge where the influence of the explosive gases is difficult to quantify. The graph of Figure 3.6 of overpressure p_s, versus scaled distance illustrates the difficulty in that there is a fairly wide spread of results at Z values of less than about 0.5.

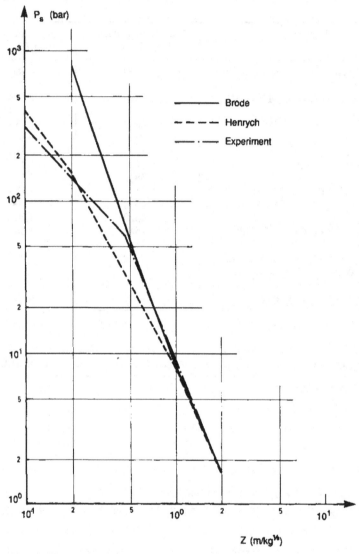

Figure 3.6 *Side-on peak overpressure vs scaled distance: comparisons between experiment and analysis*

3.5 Other important blast wave parameters

Other significant blast wave parameters include T_s, the duration of the positive phase when the pressure is in excess of ambient pressure and i_s, the specific impulse of the wave which is the area beneath the pressure-time curve from arrival at time t_a to the end of the positive phase as given by

$$i_s = \int_{t_a}^{t_a+T_s} p(t)dt \tag{3.25}$$

A typical pressure-time profile for a blast wave in free air is shown in Figure 3.7 where Δp_{min} is the greatest value of underpressure (pressure below ambient) in the negative phase of the blast of duration T^-. This is the rarefaction component of the blast wave. Brode's solution for Δp_{min} (bar) is

$$\Delta p_{min} = -\frac{0.35}{Z} \quad (Z > 1.6) \tag{3.26}$$

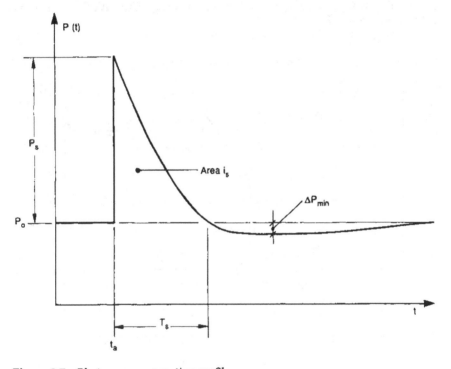

Figure 3.7 *Blast wave pressure-time profile*

The negative (or suction) phase duration is given by

$$T^- = 1.25W^{1/3} \tag{3.27}$$

and the associated specific impulse in the suction phase i^- is given by

$$i^- \approx i_s \left[1 - \frac{1}{2Z}\right] \tag{3.28}$$

Finally, λ_{rw}, the length of the rarefaction wave (based on sound speed of 340 m/s in air), is given (in metres) for T^- in seconds by

$$\lambda_{rw} = 340T^- \tag{3.29}$$

A convenient way of representing significant blast wave parameters is to plot them against scaled distance as shown in Figure 3.8, which is adapted from graphs presented in a number of references such as Baker,[8] Baker *et al.*[6] and the design code TM5-1300.[9] The variation of underpressure with scaled distance is shown in Figure 3.9.

Alternatively, parameters can be plotted against the distance from a 1 kg charge of TNT as shown in Figure 3.10, which is adapted from Kingery and Bulmash.[10] Kingery and Bulmash have curve-fitted these data which are used in the weapons effects calculation program CONWEP[11] which is based on Reference 7.

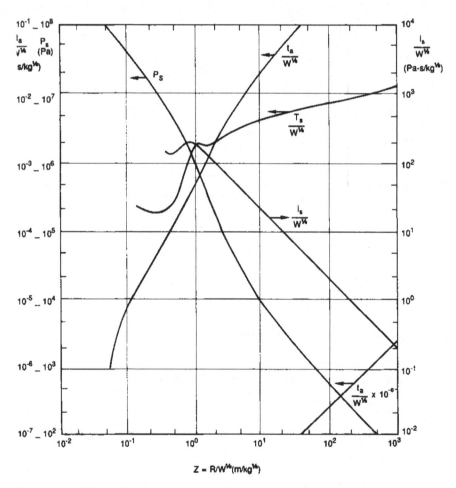

Figure 3.8   *Side-on blast wave parameters for spherical charges of TNT (after Ref. 6)*

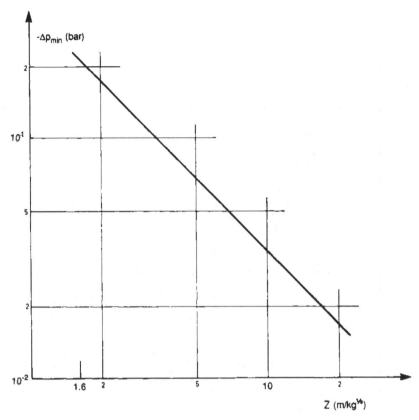

Figure 3.9 *Underpressure vs scaled distance*

The directory of significant blast wave parameters should also include dynamic pressure q_s, blast wave front speed U_s expressed as $\overline{U}$ ($= U_s/a_o$), particle velocity just behind the wavefront u_s, expressed as $\overline{u}$ ($= u_s/a_o$) and the waveform parameter b in the equation describing the pressure-time history of the blast wave (Equation 3.35 below). Figure 3.11 shows these parameters plotted against scaled distance Z.

3.6 Blast wave scaling laws

The most widely used approach to blast wave scaling is that formulated independently by Hopkinson[12] and Cranz[13]. Hopkinson–Cranz scaling is commonly described as cube-root scaling a formal statement of which (quoted by Baker *et al.*[6]) is:

> Self-similar blast waves are produced at identical scaled distances when two explosive charges of similar geometry and of the same explosive but of different sizes are detonated in the same atmosphere.

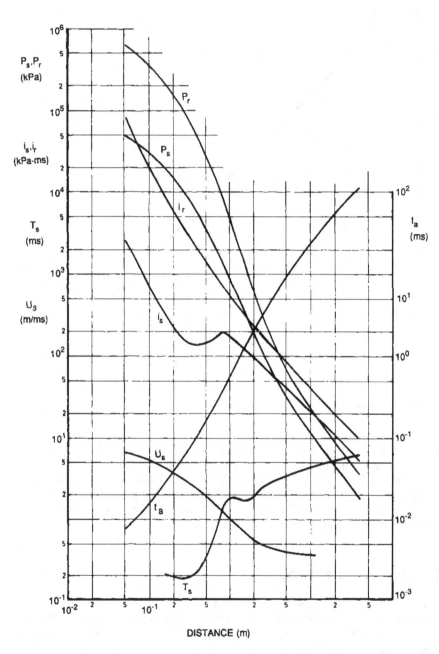

Figure 3.10   *Blast wave parameters vs distance for 1 kg TNT spherical charge (after Ref. 10)*

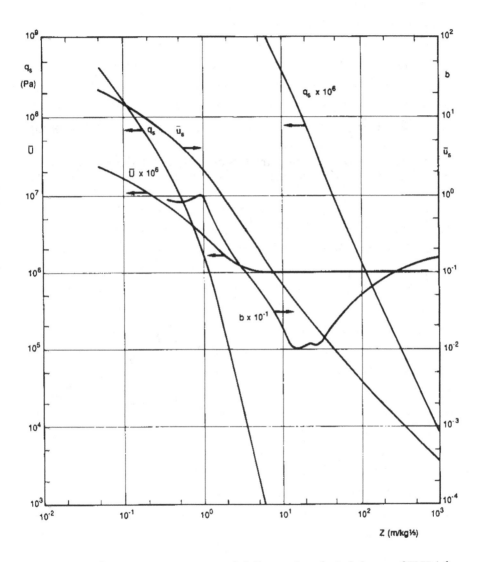

Figure 3.11 *Blast wave parameters vs scaled distance for spherical charges of TNT (after Ref. 6)*

Thus, if the two charge masses are W_1 and W_2 of diameter d_1 and d_2 respectively then, for the same explosive material, it is clear that

$$W_1 \propto d_1{}^3$$

$$W_2 \propto d_2{}^3$$

$$\therefore \quad \frac{W_1}{W_2} = \left(\frac{d_1}{d_2}\right)^3$$

$$\therefore \quad \frac{d_1}{d_2} = \left(\frac{W_1}{W_2}\right)^{1/3} \tag{3.30}$$

Therefore, if the two charge diameters are in the ratio $d_1/d_2 = \lambda$, then as Figure 3.12 indicates, if the same overpressure p_s is to be produced from the

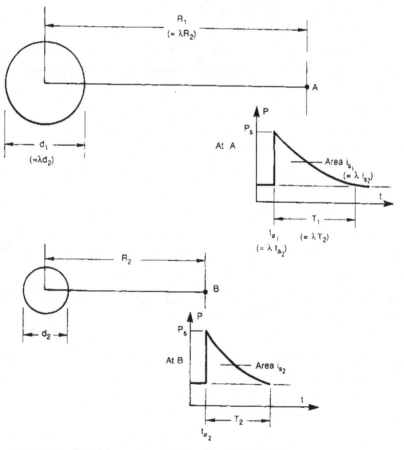

Figure 3.12   *Basis of Hopkinson–Cranz scaling law (after Ref. 6)*

two charges the ratio of the ranges at which the particular overpressure is developed will also be λ, as will the positive phase duration ratio and the impulse ratio.

Ranges at which a given overpressure is produced can thus be calculated using the results of Equation 3.30. For example

$$\frac{R_1}{R_2} = \left(\frac{W_1}{W_2}\right)^{1/3} \tag{3.31}$$

where R_1 is the range at which a given overpressure is produced by charge W_1 and R_2 is the range at which the same overpressure is generated by charge W_2.

The Hopkinson–Cranz approach leads readily to the specification of the scaled distance $Z(= R/W^{1/3})$ introduced above. It is clear that Z is in effect the constant of proportionality in relationships such as those of Equation 3.30

$$d_1 = ZW_1^{1/3}$$

$$R_1 = ZW_1^{1/3} \tag{3.32}$$

The use of Z in Figures 3.8 and 3.11 above and in 3.17 below allows a compact and efficient presentation of blast wave data for a wide range of situations.

In the case of blast waves produced from explosions at high altitude where ambient conditions are very different from those at sea level, the most commonly used scaling law is that due to Sachs[14] which states that dimensionless overpressure and impulse are unique functions of a dimensionless scaled distance. The appropriate dimensionless parameters include ambient atmospheric conditions. Thus, examples of scaled pressure, impulse and distance are respectively

$$\overline{P} = \frac{p_s}{p_0}$$

$$\overline{i} = \frac{i_s a_0}{E^{1/3} p_0^{2/3}}$$

$$\overline{R} = \frac{R p_0^{1/3}}{E^{1/3}} \tag{3.33}$$

It will be seen that, for normal sea level conditions, the Sachs scaled distance $\overline{R}$ can be reduced to the Hopkinson–Cranz scaled distance Z if explosive energy E is taken as being proportional to charge weight W.

3.7 Surface bursts

The foregoing sections refer to free-air bursts remote from any reflecting surface and are usually categorised as spherical air bursts. When attempting to quantify overpressures generated by the detonation of either high explosive sources or nuclear devices in contact with the ground, modifications must be made to charge weight or yield before using the graphs presented earlier.

Good correlation for surface bursts of condensed high explosives with free-air burst data results if an enhancement factor of 1.8 is assumed. In other words, surface bursts produce blast waves that appear to come from free-air bursts of 1.8 times the actual source energy. It should be noted that, if the ground were a perfect reflector and no energy was dissipated in producing a crater and groundshock the reflection factor would be 2.

To illustrate the situation consider the following example.

Example 3.1

Investigate the correlation between the pressures generated by the surface detonation of a 1 Mt nuclear device by making a prediction based on the free-air burst data of Figure 3.8 and the surface burst data of Figure 3.13 which gives incident overpressure plotted against positive phase duration for a range of nuclear devices.

Given that 50% of the energy of a nuclear device goes to forming the blast wave, a 1 Mt weapon equates to 500 kt of TNT. If this device is detonated as a surface burst this equates to 1.8 times this value (i.e. 0.9×10^9 kg TNT).

Now choose a specific value of overpressure on which to base a comparison. From Figure 3.13, a peak incident overpressure p_s of 100 kPa has an associated positive phase duration of 2.15 s. From Figure 3.8 this overpressure corresponds to a scaled distance Z of 2.5 and a value of scaled positive phase duration $T_s/W^{1/3}$ of 2.25×10^{-3} from which T_s is evaluated as

$$T_s = 2.25 \times 10^{-3} \times (0.9 \times 10^9)^{1/3}$$

$$= 2.17 \text{ s.} \tag{3.34}$$

This compares well with that derived from the air burst data, even though the TNT equivalent of the megatonne device is strictly outside the normal range of validity of the condensed high explosive scaled distance graphs.

Data specific to condensed high explosives detonated as hemispherical surface bursts are also available from Figure 3.14 (also adapted from Ref 10) which is presented in a similar form to that of Figure 3.10.

3.8 Blast wave pressure profiles

The pressure-time history of a blast wave is often described by exponential functions such as the Friedlander equation which has the form

$$p(t) = p_s \left[1 - \frac{t}{T_s}\right] \exp\left\{-\frac{bt}{T_s}\right\} \tag{3.35}$$

In this equation b is called the waveform parameter and is a function of peak overpressure P_s as indicated in Table 3.4 which contains values derived from the scaled distance graphs presented in Reference 6. More detailed information is also available in Kinney and Graham[15].

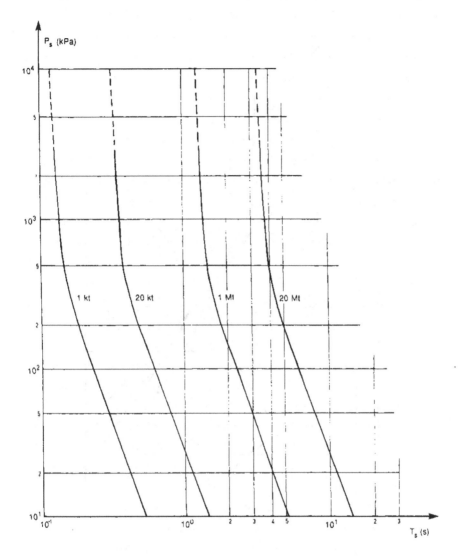

Figure 3.13 *Overpressure vs positive phase duration for nuclear surface bursts*

For many purposes, approximations are quite satisfactory. Thus, linear decay is often used in design where a conservative approach would be to represent the history by line I in Figure 3.15. Alternatively it might be desirable to preserve the same impulse in the idealised wave shape compared with the real profile as illustrated by line II in Figure 3.15 where the areas beneath the actual decay and the approximation are equal.

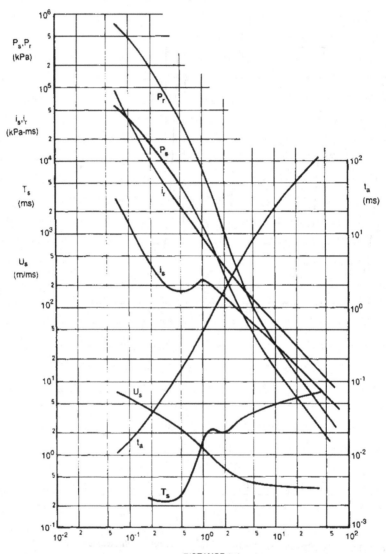

Figure 3.14   *Blast wave parameters vs distance for 1 kg TNT hemispherical surface burst (after Ref. 10)*

3.9   Blast wave interactions

When blast waves encounter a solid surface or an object made of a medium more dense than that transmitting the wave, they will reflect from it and, depending on its geometry and size, diffract around it. The simplest case is that of an infinitely large rigid wall on which the blast wave impinges at zero

Table 3.4

Z (m/kg$^{1/3}$)	b
0.4	8.50
0.6	8.60
0.8	10.00
1.0	9.00
1.5	3.50
2.0	1.90
5.0	0.65
10.0	0.20
20.0	0.12
50.0	0.24
100.0	0.50

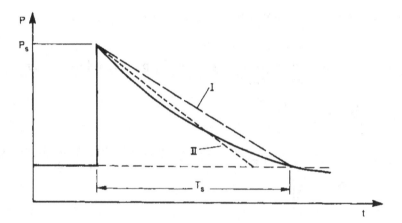

Figure 3.15 *Blast wave profiles: real, conservative, impulse equality*

angle of incidence. In this case the incident blast wavefront, travelling at velocity U_s into air at ambient pressure, undergoes reflection when the forward moving air molecules comprising the blast wave are brought to rest and further compressed inducing a reflected overpressure on the wall which is of higher magnitude than the incident overpressure. The situation is illustrated in Figure 3.16.

As mentioned above shock front parameters were first calculated by Rankine and Hugoniot derived from considerations of conservation of momentum and energy. From these equations, and assuming that air behaves as a real gas with specific heat ratio $C_p/C_v = \gamma$, all significant blast front parameters are obtainable.

For zero incidence, reflected peak pressure p_r is given by

$$p_r = 2p_s + (\gamma + 1)q_s \tag{3.36}$$

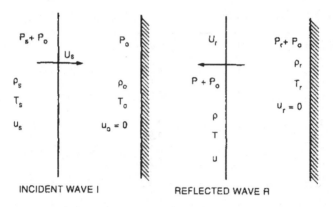

Figure 3.16   *Face-on reflection*

where the dynamic pressure q_s is

$$q_s = \tfrac{1}{2}\rho_s u_s^2 \tag{3.37}$$

Here ρ_s is the density of the air and u_s is the particle velocity behind the wavefront. It can be shown that

$$u_s = \frac{a_0 p_s}{\gamma p_o} \left[1 + \left[\frac{\gamma + 1}{2\gamma} \right] \frac{p_s}{p_o} \right]^{-\frac{1}{2}} \tag{3.38}$$

where a_0 is the speed of sound at ambient conditions. Substitution of Equations 3.37 and 3.38 into Equation 3.36 and rearrangement gives

$$p_r = 2p_s \left[\frac{7p_o + 4p_s}{7p_o + p_s} \right] \tag{3.39}$$

when, for air, γ is set equal to 1.4.

Inspection of this equation indicates that an upper and lower limit to p_r can be set. When the incident overpressure p_s is a lot less than ambient pressure (e.g. at long range from a small charge) the equation reduces to

$$p_r = 2p_s \tag{3.40}$$

When p_s is a lot greater than ambient pressure (e.g. at short range from a large charge) Equation 3.39 reduces to

$$p_r = 8p_s \tag{3.41}$$

If a reflection coefficient C_R is defined as the ratio of p_r to p_s then the Rankine-Hugoniot relationships predict that C_R will lie between 2 and 8. However, because of gas dissociation effects at very close range, measurements of C_R of up to 20 have been made. Figure 3.17 shows reflected overpressure and impulse i_r for normally reflected blast wave parameters plotted against scaled distance Z. It is worth noting that the lowest possible value of Z corresponds to the surface of the (spherical) TNT charge. If TNT is taken as being of density 1600 kg/m^3, the limiting Z value is 0.053.

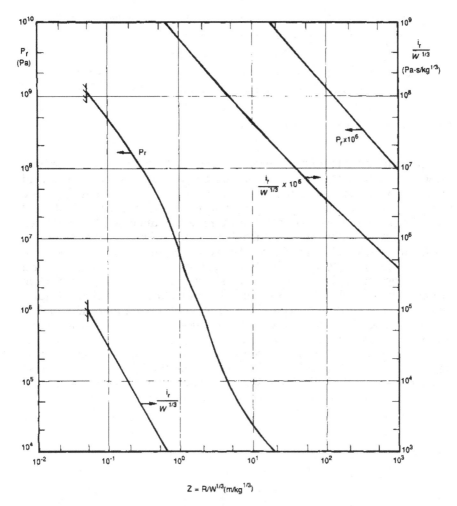

Figure 3.17 *Face-on reflection: blast wave parameters vs scaled distance for spherical charges of TNT (after Ref. 6)*

Reflected overpressure and impulse data are also given in the graphs of Figures 3.10 and 3.14 for 1 kilogramme spherical and hemispherical charges of TNT respectively.

3.9.1 *Regular and Mach reflection*

In the discussion above, the angle of incidence α_I of the blast wave on the surface of the target structure was zero. When α_I is 90° there is no reflection and the target surface is loaded by the peak overpressure which is sometimes referred to as 'side-on' pressure. For α_I between these limits either regular reflection or Mach reflection occurs.

Consider firstly regular reflection which is illustrated in Figure 3.18.

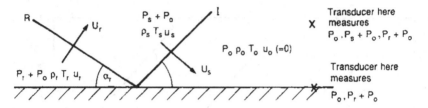

Figure 3.18   *Regular reflection (after Ref. 6)*

Important points to note are that, for a given value of p_r there is a limiting value of α_I. Above this value regular reflection does not occur and instead there is Mach reflection as described below. Also, for every gas there is a value of α_I above which p_r is greater than for p_r generated in reflection at normal incidence ($\alpha_I = 0$). For air this angle is approximately 40°. For a given value of p_s there is a value of α_I such that the ratio of p_r to p_o is a minimum. Finally the angle at which the blast wave is reflected α_r increases as α_I increases. Figure 3.19, which is of reflection coefficient C_r against α_I for a limited range of incident overpressures, shows these aspects of the reflection process. Figure 3.20 presents the same information in less detail but for a wider range of incident overpressures. It is worth noting that the Rankine-Hugoniot prediction of a maximum reflection coefficient of 8 is clearly exceeded at higher values of p_s.

The Mach reflection process occurs when α_I exceeds a threshold value dependent on incident overpressure. Mach reflection is a complex process and is sometimes described as a 'spurt'-type effect where the incident wave

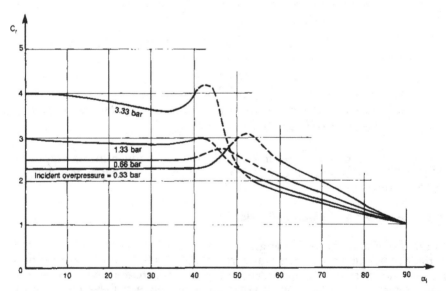

Figure 3.19   *Reflection coefficient vs angle of incidence for varying values of incident overpressure (low pressure range) (after Ref. 18)*

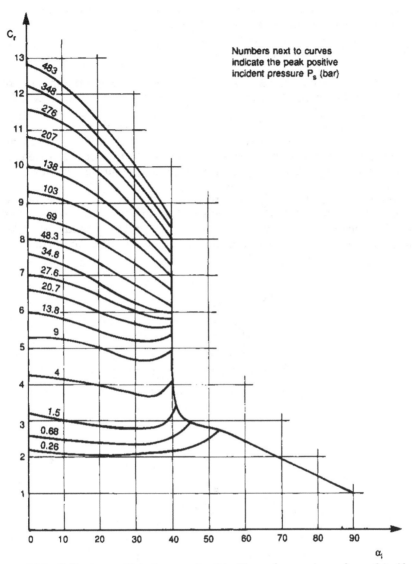

Figure 3.20 *Reflection coefficient vs angle of incidence for varying values of incident overpressure (high pressure range) (after Ref. 9)*

'skims' off the reflecting surface rather than 'bouncing' as is the case at lower values of α_I. The result of this process is that the reflected wave catches up with and fuses with the incident wave at some point above the reflecting surface to produce a third wavefront called the Mach stem. The point of coalescence of the three waves is called the triple point. In the region behind the Mach stem and reflected waves is a slipstream region where, although pressure is the same, different densities and particle velocities exist. The formation of a Mach stem is important when a conventional or nuclear device detonates at some height of burst above the ground. Mach stems also occur when a device is detonated inside a structure where the angles of incidence of the blast waves on the internal surfaces can vary over a wide range.

The situation is illustrated in Figure 3.21a which shows the three wavefronts and in Figure 3.21b which shows how the Mach stem height develops with range. Also shown is the variation of pressure on the ground from a detonation at some height of burst. It should be noted that, at the inception of Mach reflection, there is an accompanying step increase in overpressure. This resulting enhancement emphasises the significance of Mach stem formation in that, by its production, a given overpressure is generated at a greater range than for regular reflection.

The importance of Mach stem formation in the context of weapons detonated at some height of burst is emphasised in Figure 3.22 which shows pressure levels (given in kiloPascals) produced on the ground by a 1 kilotonne nuclear device initiated at different heights of burst. To illustrate this consider the following example.

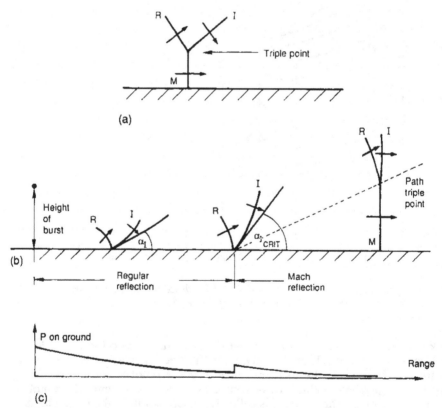

Figure 3.21   *(a) Mach stem triple point formation; (b) Mach stem and triple point development for height of burst explosion; (c) Pressure on ground vs range*

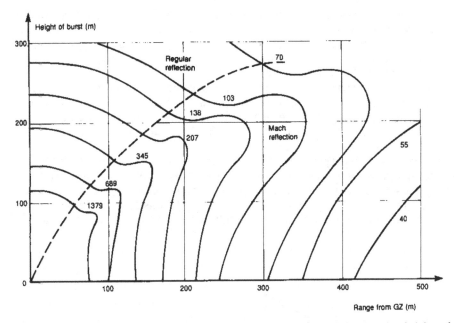

Figure 3.22 *Overpressure contours for 1 kt nuclear device detonated at varying heights of burst with range from ground zero (after Ref. 18)*

Example 3.2
For successful attack on a target it is necessary to produce an overpressure of 103 kPa at a range just in excess of 350 m from a 1 kilotonne device. Inspection of Figure 3.22 indicates that if the device were to be initiated at a height of burst of 200 m, over-pressures in excess of 103 kPa will be generated out to a range of 355 m (the greatest possible for the given overpressure). This height of burst is referred to as the optimum height of burst for that particular pressure and device. Any height of burst lying below the broken line in Figure 3.22 will produce Mach reflection and consequent enhancement of range for a given overpressure. Above the line, regular reflection occurs and the range at which a given overpressure is generated will be less. It is also worth noting that if the weapon were initiated on the ground, although pressures in excess of 103 kPa would be generated this would only occur out to a range of about 245 m from ground zero (the point on the ground vertically below the centre of the device) because energy will be transmitted to the ground in producing cratering and groundshock waves (see below).

3.10 Blast wave external loading on structures

The foregoing discussion is centred on reflecting surfaces that are essentially infinite and do not allow the diffraction process described below to occur. In

the case of finite target structures three classes of blast wave structure interaction can be identified.

The first of these is the interaction of a large-scale blast wave such as might
be produced by a nuclear device of substantial yield. The situation is illustrated schematically in Figure 3.23a.

Here the target structure is engulfed and crushed by the blast wave. There
will also be a translational force tending to move the structure bodily later-
ally (a drag force) but because of the size and nature of the structure it is
unlikely actually to be moved. This is diffraction loading and the target is
described as a diffraction target.

The second category is where a large-scale blast wave interacts with a
small structure such as a vehicle. Here the target will again be engulfed
and crushed. There will be a more or less equal 'squashing' overpressure
acting on all parts of the target and any resultant translatory force will only
last for a short time. However, more significantly, a translational force due to

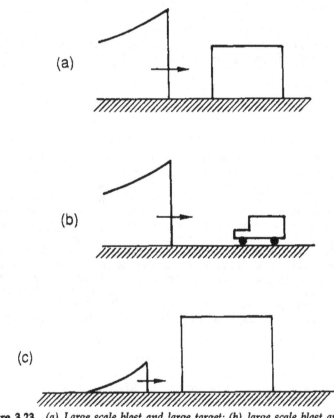

Figure 3.23   (a) *Large scale blast and large target; (b) large scale blast and small target;
(c) small scale blast and large target*

dynamic or drag loading will act for sufficiently long to move the target and it is likely that a substantial part of the resulting damage will be as a consequence of this motion. The target structure is described as a drag target, as shown in Figure 3.23b. The load-time history is described in more detail below.

Finally, consider the case of a blast wave produced by the detonation of a smaller charge loading a substantial structure as shown in Figure 3.23c. Here, the response of individual elements of the structure needs to be analysed separately since the components are likely to be loaded sequentially.

For the first and second situations above, consider the load profile for each structure. Each experiences two simultaneous components of load. The diffraction of the blast around the structure will engulf the target and cause a normal squashing force on every exposed surface. The structure experiences a push to the right as the left hand side of the structure is loaded followed closely by a slightly lower intensity push to the right as diffraction is completed. The drag loading component causes a push on the left side of the structure followed by a suction force on the right and side as the blast wave dynamic pressure (the blast wind) passes over and round the target. The situation is illustrated schematically in Figure 3.24 which also shows the location of the wavefront on the top and sides of the target at the times indicated.

With reference to Figure 3.25 which shows the 'squashing' and dynamic pressure variation at significant times on the structure, the following points should be noted.

In Figure 3.25a the peak pressure experienced by the front face of the target at time t_2 will be the peak reflected overpressure p_r. This pressure will then decay in the time interval $(t' - t_2)$ because the pressure of the blast wave passing over the top of the structure and round the sides is less than p_r (the peak top and side overpressure will be p_s). Thus, decay in front face overpressure continues until the pressure is equal to the stagnation pressure $p_{stag}(t)$ given by

$$p_{stag}(t) = p_s(t) + q_s(t) \tag{3.42}$$

where $q_s(t)$ is the blast wave dynamic pressure at time t. The time t' is given approximately by

$$t' = \frac{3 \times S}{U_s} \tag{3.43}$$

where S is the smaller of B/2 or H as defined in Figure 3.26 below and U_s is the blast front velocity.

After time t' the front face pressure will be a time-varying function of the static and dynamic overpressure components of the wave. In Figure 3.25b the deviation from the linear decay of pressure on top and sides after time t_3 is due to the complex vortices formed at the intersection of the top and the sides with the front. In Figure 3.25c the load profile on the rear face is of finite rise time because of the time required by the blast wave to travel down the rear of the target to complete the diffraction process. Figures 3.25d and

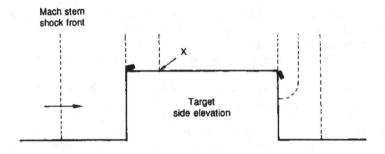

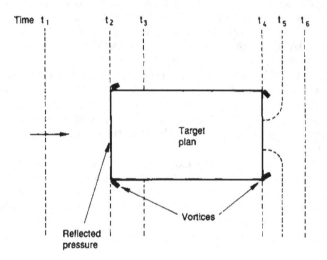

Figure 3.24 *Diffraction of blast wave round simple geometry target*

3.25e show the forces exerted on the front and rear faces of the target by the dynamic, drag or wind forces. Drag force F_D is given by

$$F_D = C_D \times q_s(t) \times A \tag{3.44}$$

where A is the area loaded by the pressure and C_D is the drag coefficient of the target, a factor that depends on target geometry.

Combining the loading from both diffraction and drag components gives the overall translatory force-time profile as shown in Figure 3.27.

If the target is relatively small (having only short sides) the interval $(t_4 - t_2)$ is small and area I in Figure 3.27 is small while area II is proportionately bigger. This loading is characteristic of a drag target.

To distinguish between diffraction and drag loading quantitatively consider the following example.

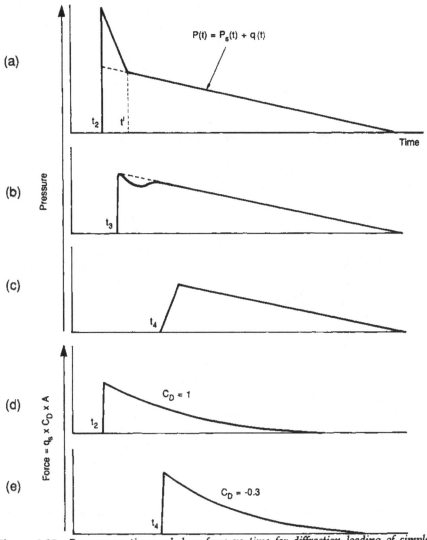

Figure 3.25 *Pressure vs time and drag force vs time for diffraction loading of simple geometry target*

Example 3.3.
A 1 kilotonne nuclear weapon is equivalent to 500 000 kg TNT. Considering a target structure at 250 m range the scaled distance Z is evaluated as

$$Z = \frac{250}{(5 \times 10^5)^{1/3}} \approx 3 \tag{3.45}$$

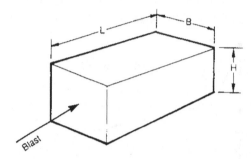

Figure 3.26 *Dimensions of simple geometry target (after Ref. 18)*

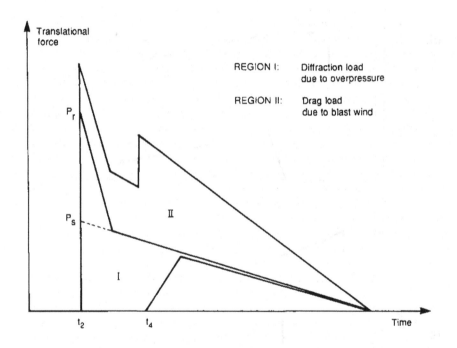

Figure 3.27 *Translational force vs time for diffraction of blast wave round simple geometry target*

From Figure 3.8 the side-on pressure at this scaled distance is 70 kPa and using Figure 3.11 the wavefront speed U_s is 440 m/s assuming that the ambient speed of sound is 350 m/s. If the target is a small military structure, say 7 m long, then the time for the blast wave to traverse the target is

$$t_2 \to t_4 = \frac{7}{440} \times 1000 \text{ ms} \approx 16 \text{ ms} \tag{3.46}$$

The positive phase duration of this blast wave is obtained from Figure 3.8 as

$$\frac{T_s}{W^{1/3}} = 2.5 \times 10^{-3} \tag{3.47}$$

from which T_s is approximately 200 ms.

Compare this with the performance of a 500 kg charge of TNT at a range of 25 m. Here the same overpressure is generated since Z is still 3 as above. However, now the positive phase duration is given by

$$T_s = 2.5 \times 10^{-3} \times 500^{1/3} = 0.02\,\text{s} = 20\,\text{ms} \tag{3.48}$$

Thus, in the case of a nuclear blast with a long positive phase duration, dynamic or drag loading lasts for all of positive phase: the structure is a drag target for this sort of loading. However, if for conventional explosives T_s is short or if the target structure is long (thus not allowing pressure equalisation between front and back faces) the target becomes more a diffraction target.

3.11 Examples of blast effects on structures

Mention will be made here of situations where structures have been subjected to blast loads from such sources as terrorist devices and large-scale nuclear weapons. In the case of terrorist bombings, there are many examples perpetrated by terrorist organisations in Northern Ireland, the Middle East and elsewhere. In the area of nuclear loading there is information about structural behaviour from the Hiroshima and Nagasaki bombs of the Second World War and the subsequent simulations of such events carried out using large conventional explosive charges.

3.11.1 Terrorist attacks

The range of weapons used by terrorists extends from a few grammes of explosive material contained in a letter bomb up to several thousand kilogrammes of material used in a vehicle bomb. Elliott[16] describes a number of such attacks while Elliott *et al.*[17] describe some general features of terrorist events.

At the top end of the range of possible attack scenarios is the case of the bombing of the US Marine Corps Battalion Headquarters Beirut in 1983 which involved a vehicle bomb containing approximately 5 tonnes of explosive material. The substantial reinforced concrete frame building was completely demolished on explosion of the device which was driven at speed at the structure and embedded itself in the front of the building. Over 240 people were killed and more than 60 wounded by this attack. A further example from the Middle East involving approximately 2.5 tonnes of explosive material was the attack on the US Embassy Annex at Antelias, East Beirut, in 1984. Here the bomber attempted to detonate the bomb in the Annex underground car park but was prevented from so doing by prompt action by soldiers guarding the building. In the event, detonation occurred on the sunken approach road to the car park at some 7 to 8 metres from the building. A retaining wall provided shielding of the structure and, coupled with the (relatively small) stand-off, the casualty level was low with only 11 deaths recorded.

A graphic example of the results of even partial confinement of blast waves is provided by a Provisional IRA attack in Belfast in 1985. A 200 kg car bomb was parked in an alley. On one side of the alley was a modern reinforced concrete frame building, on the other a relatively old masonry building used as offices by a firm of solicitors. When the bomb went off the masonry structure was completely destroyed, the reinforced concrete structure was damaged (but repairable) and the confining effect of the alley produced damaging blast waves up to five or six storeys high causing extensive damage to glazing in a modern office block on the far side of the street.

There are many other examples of car bomb attacks, particularly in the Middle East. Instances where a car or lorry bomb has gained (sometimes forcibly) entrance to a structure include the Iraqi Embassy in Beirut, where in December 1981 an explosion in the embassy underground garage caused 20 deaths. In November 1982 at the Israeli Military HQ in Tyre a massive explosion in contact with the eight storey building killed 89 people. On the same day as the US Marines HQ was attacked, the French Paratroop Command Post in Beirut suffered a huge explosion in contact with the nine-storey building and 58 people were killed.

Other examples where car bombs have been activated near structures and have caused serious effects include the National Palace, Guatemala City, in September 1980 which caused extensive damage and killed eight. In Teheran in October 1981 a car bomb in a main square demolished a five storey hotel and severely damaged other buildings: 60 were killed. At the South African Air Force HQ in Pretoria in May 1983 a car bomb detonated adjacent to the building and 19 were killed. Closer to home, at the Harrods Store in London just before Christmas in 1983, a 12 kg explosive device detonated and six were killed. There have also been terrorist bombings in the heart of the City of London in the spring of 1992 when extensive structural damage was caused to a number of buildings including the Baltic Exchange. On the same day a vehicle bomb rendered unsafe the Staples Corner flyover in North London. In 1993 a massive device once again brought the City to a temporary standstill causing damage over a wide area.

There are many instances where a package bomb exploding inside a building has caused extensive damage to the structure together with death and injury. Some examples include the telephone exchange in Madrid where in May 1982 an 80 kg device severely damaged the building cutting all telecommunications with Madrid and caused about £7m damage. Also in Spain at a US Air Base near Madrid in April 1985 a bomb detonated near a restaurant demolishing it, killing 18 and injuring 79. At Government Offices in Bastia, Corsica, in March 1987 five 10 kg charges damaged the building and destroyed computers and records causing about £6m damage.

Early in 1991 the Provisional IRA first used home-made mortars on main-land Britain with an attack on Downing Street timed to coincide with a

Cabinet meeting. Fortunately, members of the Government were unhurt even though one of the mortars exploded in the garden at the rear of the building only a few metres from the Cabinet Room.

3.11.2 Nuclear attacks

The bombing of Hiroshima and Nagasaki on the 5th and 9th August 1945 was accomplished using devices equivalent to approximately 12.5 and 22 kilotonnes detonated at heights of burst of 513 m and 505 m respectively. Examples of the effects on a wide variety of target structures are available from photographs[18] taken during the days following the Japanese surrender on 14th August 1945. Structural damage extended over a wide range with failure of buildings and building elements occurring out to distances in excess of 2 km from ground zero.

3.11.3 Simulations

The limited nuclear test ban treaty of the early 1960s meant that atmospheric nuclear testing in the West ceased. However, the effects of large-scale blast waves on structures has remained the subject of considerable research worldwide. The biennial series of nuclear blast simulations in the USA has allowed such studies to continue. Most recent events are Direct Course (1983), a height of burst trial involving 600 t of ammonium nitrate and fuel oil (ANFO) held in a spherical container on top of a 50 m high tower, and hemispherical surface bursts using up to 4500 t of ANFO such as Minor Scale (1985), Misty Picture (1987), Miser's Gold (1989) and Distant Image (1991). These trials are well documented in the proceedings of the so-called 'MABS' symposia.[19]

The use of blast tunnels for structural response assessment in a nuclear blast environment is a feature of work in the UK, France, Germany and Canada. In the United Kingdom the Atomic Weapons Establishment at Foulness operates a blast tunnel driven by the detonation of a relatively small amount of conventional explosive. Overpressures of about 45 kPa with a positive phase duration of up to about 300 ms are achievable in the 10.7 m diameter working section. A similar facility exists at the Centre d'Etudes de Gramat near Toulouse in France. This tunnel is driven by compressed gas initially contained in an array of seven pressurised cylinders. Explosive cutting of gas-tight diaphragms releases the compressed air to produce the required blast simulation. The most recent European facility to be commissioned is at Reiteralpe in Germany and makes use of mountainside tunnels in which to generate simulated nuclear blast by the use of compressed air.

It should also be noted that there are a number of open-ended shock tubes throughout the world offering simulation of blast waves produced by the firing of a gun up to larger facilities allowing investigation of the effects of several kilogrammes of high explosive material. There is also, of course, a substantial amount of work worldwide using small-scale charges for assessment of the effects of conventional explosives on structures. Into this category fall the testing by government agencies of protective structures to counter terrorist attack and investigations of accidental explosions in ammunition storage facilities.

3.12   References

[1] Brode H.L., Numerical solution of spherical blast waves. *Journal of Applied Physics* No. 6 June 1955.

[2] Kingery C.N., Airblast Parameters versus Distance for Hemispherical TNT Surface Bursts. BRL Report No. 1344 Aberdeen Proving Ground Maryland USA (1966)

[3] Henrych J., *The Dynamics of Explosion and Its Use*. Elsevier, Amsterdam (1979)

[4] Rankine W.J.H., *Philosophical Transactions of the Royal Society*. V160, 277 (1870)

[5] Liepmann H.W., Roshko A., *Elements of Gas Dynamics*. Wiley, New York (1957)

[6] Baker W.E., Cox P.A., Westine P.S., Kulesz J.J., Strehlow R.A., *Explosion Hazards and Evaluation*. Elsevier, Amsterdam (1983)

[7] TM5-855-1, *Fundamentals of Protective Design for Conventional Weapons*. US Dept. of the Army Technical Manual (1987)

[8] Baker W.E., *Explosions in Air*. University of Texas Press, Austin, Texas (1973)

[9] TM5-1300, *Design of Structures to Resist the Effects of Accidental Explosions*. US Dept. of the Army Technical Manual (1991)

[10] Kingery C.N., Bulmash G., *Airblast Parameters from TNT Spherical Air Burst and Hemispherical Surface Burst*. US Army Armament Research and Development Center, Ballistics Research Lab., Aberdeen Proving Ground, Maryland. Tech. Report ARBRL-TR-02555 (1984)

[11] CONWEP: Conventional Weapons Effects Program. Prepared by D.W. Hyde, US Army Waterways Experimental Station, Vicksburg (1991)

[12] Hopkinson B., British Ordnance Board Minutes 13565 (1915)

[13] Cranz C., *Lehrbuch der Ballistik*. Springer-Verlag, Berlin (1926)

[14] Sachs R.G., *The Dependence of Blast on Ambient Pressure and Temperature*. BRL Report 466, Aberdeen Proving Ground, Maryland, USA (1944)

[15] Kinney G.F., Graham K.J., *Explosive Shocks in Air*. (2nd edition) Springer-Verlag, New York (1985)

[16] Elliott C.L., The defence of buildings against terrorism and disorder – a design philosophy for the construction of ordinary buildings and installations to resist terrorism and public disorder. M.Phil. Thesis, Cranfield Institute of Technology (1990)

[17] Elliott C.L., Mays G.C., Smith P.D., The Protection of Buildings against Terrorism and Disorder. *Proceedings of the Institution of Civil Engineers. Structures and Buildings*, **94**, August, 287–97 (1992)

[18] Glasstone S., Dolan P.J., *The Effects of Nuclear Weapons*. United States Departments of Defense and Energy (Third Edition) (1977)

[19] Military Applications of Blast Simulation Symposia Proceedings. No. 6 Cahors, France (1979), No. 7 Medicine Hat, Canada (1981), No. 8 Spiez, Switzerland (1983), No. 9 Oxford, UK (1985), No. 10 Bad Reichenhall, Germany (1987), No. 11 Albuquerque, USA (1989), No. 12 Perpignan, France (1991), No. 13 The Hague, Netherlands (1993)

Symbols

a_0 velocity of sound at ambient conditions

a_{ex} velocity of sound in unreacted explosive

a_g	velocity of sound in explosive gases
A	area of target loaded by blast
b	wavefront parameter
B	target dimensions
C_R	reflection coefficient
C_D	drag coefficient
C_p	specific heat at constant pressure
C_v	specific heat at constant volume
D	detonation wave speed
d	charge diameter
E	total explosive energy
e_o	specific energy of atmosphere
e_{ex}	specific energy of explosive at ambient conditions
F_D	drag force
H	target dimensions
i_s	specific side-on impulse
i_r	specific reflected impulse
$\bar{i}$	scaled impulse
$\bar{i}^-$	negative phase specific impulse
$\bar{p}$	scaled pressure
p	pressure
p_{ex}	ambient pressure of explosive
p_o	atmospheric pressure
p_r	peak reflected overpressure
p_s	peak side-on overpressure
p_{stag}	stagnation pressure
Δp_{min}	peak underpressure
p_1	pressure at detonation wavefront
q_s	peak dynamic pressure
Q	mass specific energy of condensed high explosive
dQ	increment of heat
r	radial distance from charge centre
R	range from charge centre
$\bar{R}$	scaled range
R_c	radius of spherical charge
R_d	radius of detonation wavefront
R_f	radius of blast wavefront
R_g	radius of explosive gases
R	gas constant
S	target dimensions
dS	change in entropy
t	time
t'	pressure reduction time
t_a	arrival time of blast wavefront
T_s	positive phase duration
T^-	negative phase duration
T	temperature
T_{ex}	ambient temperature of explosive
T_o	atmospheric temperature

u	particle velocity
u_d	particle velocity behind detonation wavefront
u_g	particle velocity of explosive gases
u_s	particle velocity behind blast wavefront
U_s	blast wavefront speed
$\bar{u}$	Mach number of wavefront particle velocity
$\bar{U}$	Mach number of wavefront
v	specific volume
W	mass of spherical TNT charge
Z	scaled distance
α_I	angle of incidence
α_r	angle of reflection
γ	specific heat ratio
λ	scale factor
λ_{rw}	length of rarefaction wave
ρ	density
ρ_g	density of explosive gases
ρ_m	density of surrounding medium
ρ_s	density behind blast wavefront
ρ_1	density behind detonation wavefront

4 Internal blast loading

4.1 Flame propagation in gas-air mixtures

Whereas a condensed high explosive undergoes a detonation reaction, when a gas-air mixture explodes the process is generally one of very rapid burning or deflagration. The speed with which a flame front propagates through a gas-air or inflammable vapour-air mixture during an explosion determines the rate at which burnt gases are generated. As an aid to identifying the regions of relevance in such an event consider Figure 4.1 which distinguishes between a stationary pre-mixed flame such as might be developed in a gas burning device and a propagating or explosion flame such as might be formed by the accidental ignition of a gas-air mixture.

The speed with which the thin reaction zone or flame front moves through the gas-air mix, measured with respect to some fixed position, is called the flame speed S_f and is related to a property of the mixture called burning velocity S_o, the velocity with which the flame front moves relative to the unburnt mixture immediately ahead of it. These two quantities are, in gen-

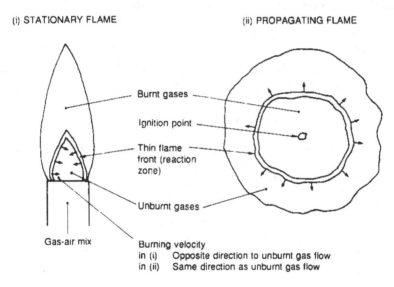

(i) STATIONARY FLAME   (ii) PROPAGATING FLAME

Burnt gases

Ignition point

Thin flame front (reaction zone)

Unburnt gases

Gas-air mix

Burning velocity
in (i)   Opposite direction to unburnt gas flow
in (ii)   Same direction as unburnt gas flow

Figure 4.1 *Stationary and propagating flames from gas-air mixtures (after Ref. 1)*

eral, not the same since, during combustion, the flame front is often also pushed forward by the effect of the expansion of gases trapped behind it.

The following analysis, adapted from Harris,[4] shows how S_f and S_o are related. If the rate of production of burnt mass m_b, dm_b/dt, is written as

$$\frac{dm_b}{dt} = A_f \rho_u S_o \tag{4.1}$$

where A_f is flame area and ρ_u is the density of unburnt gas. This equation may be rewritten as

$$\rho_b \frac{dV_b}{dt} + V_b \frac{d\rho_b}{dt} = A_f \rho_u S_o \tag{4.2}$$

using the fact that $m_b = \rho_b V_b$ where ρ_b and V_b are the density and volume of the burnt gas respectively. Assuming that r is the distance travelled by the flame from the point of ignition and that $A_f = A_n$ (where A_n is the minimum area possible for the flame to adopt normal to the direction of flame propagation) then we have

$$\frac{dV_b}{dr} = A_n = A_f \tag{4.3}$$

Using this equation and noting that $dr/dt = S_f$ then

$$S_f = \left(\frac{\rho_u}{\rho_b}\right) S_o - \left(\frac{V_b}{\rho_b A_f}\right)\left(\frac{d\rho_b}{dt}\right) \tag{4.4}$$

For situations where the flame front area is complex (ie A_f and A_n are unequal), then a more general relationship is

$$S_f = \left(\frac{A_f}{A_n}\right)\left(\frac{\rho_u}{\rho_b}\right) S_o - \left(\frac{V_b}{\rho_b A_n}\right)\left(\frac{d\rho_b}{dt}\right) \tag{4.5}$$

It should be noted, however, that this derivation assumes that before ignition the gases are at rest. In reality gas movement will also contribute either to increase or decrease flame speeds.

In many cases the relationship between flame speed and burning velocity can be simplified with only a small loss in accuracy. For example, in the case of a planar wave travelling down a tube or duct or a smooth hemispherical or spherical flame propagating freely and with the assumption that gases are initially at rest and that the flame is laminar, the relationship reduces to

$$S_f = S_o E \tag{4.6}$$

where E is the expansion factor, a measure of the increase in volume produced by combustion and given by the ratio of the densities of unburnt gas to burnt gas ($= \rho_u/\rho_b$). Expansion ratio can also be written as

$$E = \left(\frac{T_f}{T_i}\right)\left(\frac{N_b}{N_u}\right) \tag{4.7}$$

where T_f is the temperature to which the gases are raised in combustion, T_i is the initial gas temperature, N_b is the number of moles of products and N_u is the number of moles of reactants. Typically, for the combustion of a hydrocarbon-air mixture, the mole number ratio is approximately unity. Consider the reaction

$$C_5H_{12} + 8(O_2 + 3.76N_2) = 6H_2O + 5CO_2 + 30.08N_2 \qquad (4.8)$$

which relates to the combustion of pentane in air. Here the number of reactant moles N_u is 39.08 and the number of product moles 41.08. Hence, to a reasonable approximation, E is given by the ratio of flame temperature to initial gas temperature.

4.2 The effect of flame area

The foregoing description of flame speeds and burning rates applies to the idealised conditions specified. The predictions take no account of any potential increase in the volumetric rate of combustion caused by increases in flame area. A more accurate relationship between flame speed and burning velocity is therefore given by

$$S_f = S_o E \left(\frac{A_f}{A_n} \right) \qquad (4.9)$$

where the term in brackets is an area correction term. This is needed because in any unit time interval the volume production of burnt gases which expand to drive the flame forward is determined by the product of the burning velocity and the actual surface area of the flame in contact with the unburnt mixture. A simple idealised example is shown in Figure 4.2 where a plane

(a) Plane flame, Area A

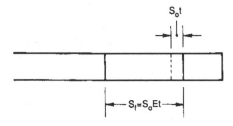

(b) Hemispherical flame, Area 2A

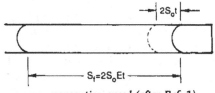

Figure 4.2 *Effect of flame area on propagation speed (after Ref. 1)*

flame front is contrasted with a hemispherical flame front as the flames propagate along a duct.

The reality of the situation may be that an even greater elongation of flame front is apparent due to the effect of the confining walls.

Thus, one should be aware that correcting for flame front area may be difficult because of the problems of defining A_f. For a spherical flame front, even after a few centimetres of expansion, the flame surface can exhibit peaks and troughs (often collectively called 'wrinkles') that increase the value of A_f over that given by the macroscopic measure of flame frontal area, $4\pi r^2$.

Very important in the determination of resultant burn velocities and their increase above the laminar burn velocities that have been discussed so far is the action of turbulence. This turbulence might be generated by the flow of unburnt gases ahead of the flame, over and around obstacles within an enclosure, during the course of an explosion. During turbulent burning the value of the burn velocity is increased by the increased transport of heat and mass within the flame front caused by random movements of eddies within the turbulent flow. Quantification of the effects of turbulence and specifying values of turbulent burning velocity is difficult because they are influenced by both the scale and intensity of the turbulent flow.

Attempts to introduce the effects of turbulence on burning velocity can be done by the introduction of a turbulence factor β so that

$$S_T = \beta S_o \tag{4.10}$$

where S_T is turbulent burning velocity. Specifying the value of β is not easy and remains rather subjective. Depending on particular circumstances, the factor ranges from about 1.5 up to 5 though in some situations values of between 8 and 10 have been proposed.

4.3   Detonation

4.3.1   Turbulence increase.

The mode of flame propagation described above is usually referred to as deflagration where the flame speed is limited by transport processes, such as heat and mass transfer which affect the fundamental burning velocity. Under some circumstances, however, flames can travel much faster, at a velocity in excess of the local speed of sound. In such cases the mode of flame propagation is called a detonation. In a detonation, as described above for condensed explosives, reaction in the gas-air or vapour-air mixture is caused by compression and heating of the gases following the passage of an intense shock wave. The shock wave and the flame front are coupled together and travel through the gas-air mixture at a constant velocity of about 1800 m/s. The pressure profile in a detonation process is characterised by a very short rise time and can produce a very high peak overpressure as has been discussed in the context of condensed high explosives.

A possible scenario for such an event could be an accident on an offshore gas production platform where a leakage has occurred. If the gas-air mixture has been ignited the flame will initially develop very much as in the case of a 'confined' explosion (see below). However, compression waves so produced will radiate outwards and cause an acceleration of the flow ahead of the flame. The outward flow velocities will increase with increasing flame area and burning velocity. In this particular scenario, obstacles in the way of the flame front (for example pipework on the platform) can induce 'flamefolds' and, coupled with the turbulence so generated, the flame area and velocity increase still further. This leads to an increase in volume production rate and this in turn leads to increased flame folding and turbulence as a sort of 'feedback' system is developed. The net result is an exponential increase in flame velocity to a level where it is in excess of the local speed of sound. If the compression waves generated by the deflagration process are relatively high, chemical energy density in the gas is intensified as is energy release and, as a consequence, the flame undergoes transition and superposition and the result is a shock or blast wave as demonstrated in Figure 4.3a which shows the overpressure record obtained during a 'congested-volume' test involving a pipe-rack on an offshore structure.

An interesting paradox is evident in this situation. Whereas here flame acceleration is enhanced by obstacles and is aided by venting from the perimeter of the volume, this is in contrast to the effects in, say, the deton-ation of a condensed high explosive in confinement where venting leads to reduced pressures.

4.3.2 Booster charges

In a military context, gas-air mixtures that are caused to detonate are known as fuel-air explosives (FAEs).

As an example, consider a requirement to cover a target zone with explos-ive material that could be detonated to produce a blast overpressure sig-nature over a wide area. Such deployment might be necessary for mine clearance: the production of a substantial blast overpressure impulse could be sufficient to initiate shallow-buried devices.

An FAE system for accomplishing this could contain a volatile liquid such as ethylene oxide or propylene oxide held in a frangible canister. The canister could be burst by the initiation of a small, condensed, high explosive charge inside causing a cloud of vaporised fuel to be dispersed and mixed with air to produce an extensive fuel-air cloud. After a suitable delay this cloud could then be detonated by means of a small booster charge of high explosive to produce a long duration blast overpressure signature such as that shown in Figure 4.3b.

4.4 TNT equivalence

Just as it is attractive to be able to compare the performance of different condensed high explosive materials on the basis of their TNT equivalence, this is similarly the case with gas and vapour cloud explosions. In order to obtain a TNT equivalence it is recommended that the combustion energy of a flammable cloud be multiplied by an efficiency factor of about 0.04 if

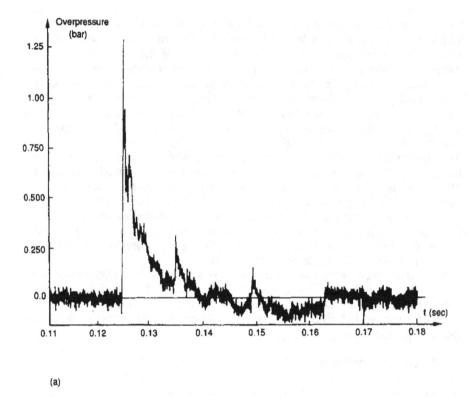

(a)

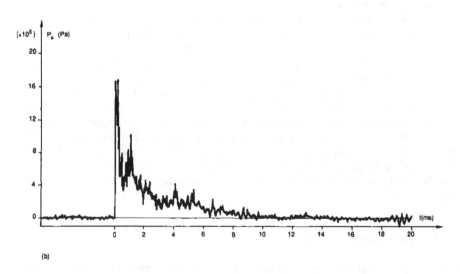

(b)

Figure 4.3   *(a) Overpressure recorded during experiment on model offshore structure pipe-rack (after Ref. 13); (b) pressure vs time for a fuel-air explosive device (after Ref. 10)*

Table 4.1

Material	Energy ratio	Efficiency factor	TNT equivalence
Hydrocarbons	10	0.04	0.4
Ethylene oxide	6	0.10	0.6
Vinyl chloride Monomer	4.2	0.04	0.16
Acetylene oxide	6.9	0.06	0.4

measurements are made from the cloud centre. Since the energy of combustion of a hydrocarbon is about 10 times the energy of detonation of TNT then 1 tonne of a hydrocarbon is therefore equivalent to about 0.4 tonnes of TNT. Some suggested values for conversion factors are given in Table 4.1 adapted from the Institution of Chemical Engineers Hazard Assessment Panel.[2]

Here 'Energy ratio' is the ratio of the energy of combustion to the energy of detonation of TNT. For other flammable gases it is suggested that the same efficiency factor as for hydrocarbons be used.

The advantages of such an approach are that there are a lot of data related to damage to structures associated with a TNT yield. The disadvantages of doing this are that predictions of overpressures are likely to be unrealistic in the near field and the precise value of efficiency factor is difficult to determine.

4.5 Internal blast loading of structures

When an explosion occurs within a structure it is possible to describe the structure as either *unvented* or *vented*. An unvented structure is likely to be stronger to resist a particular explosive event than a vented structure where some form of pressure relief is incorporated in the design. In this section consideration will be given not only to the detonation of condensed high explosive material inside a building but also the overpressures generated by gas and dust explosions within structures.

4.5.1 Internal loading by condensed high explosives

The detonation of a condensed high explosive inside a structure produces two loading phases. Firstly, reflected blast overpressure is generated and, because of the confinement provided by the structure, re-reflection will occur and (depending on the geometry of the confining structure) Mach stems could be formed. The reflection process will produce a train of blast waves of decaying amplitude. While this is happening the second phase develops as the gaseous products of detonation independently cause a build-up of pressure. This component of loading is called the 'gas pressure' or quasi-static pressure. The load profile for the structure will thus be very complex.

It is fairly straightforward to estimate the magnitude of the reflected blast wave parameters (p_r, i_r) by using, for example, the scaled distance curves of

Figure 3.17. Quantification of the magnitude of re-reflected waves is generally more difficult though for symmetrical structures predictions are possible and reasonably accurate. For example, analyses exist for spherical containment structures with a centrally detonated charge and involve relatively few simplifying assumptions as presented by Baker *et al.*[3] Loading of cylindrical vessels can also be assessed reasonably accurately as evidenced by Figure 4.4 which show a comparison between experimental measurements and numerical prediction for a cylindrical containment structure.

Prediction is more difficult in the case of oblique reflections inducing Mach stem waves and the complex pressure enhancements that occur at internal corners in box-like enclosures. However, programs such as the BLASTIN code[4] allows investigation of the multiple shock wave reflections from the floor, walls and roof produced by conventional high explosive detonations in closed rectangular box-shaped rooms together with the subsequent gas pressure phase. The BLASTIN program finds application in the situation illustrated schematically in the Figure 4.5

A specific example is provided in Figure 4.6 where the complex pressure-time history at the centre of a wall in a $10 \times 10 \times 10$ m room in which a 5 kg charge of TNT detonated in the centre of the floor is shown. The first part of the record relates to the reverberating blast wave phase while the second part shows the gas pressure phase. Since this structure is effectively unvented and BLASTIN has no provision for the cooling of these gases (which would reduce internal pressure) this pressure persists indefinitely.

Equally complex is the prediction of blast load profiles within a room for detonation external to the structure. The CHAMBER code[5] was developed to compute the pressure in a rectangular box-shaped room produced by the airblast from external explosions penetrating into the room through openings in the walls such as doors, windows and ducts of air entrainment systems. The situation is illustrated in Figure 4.7.

The graph of Figure 4.8 illustrates the pressure-time history within a $10 \times 10 \times 10$ m room containing a single 2×2 m opening in one wall at the middle of the wall opposite the opening when a GP250 bomb (containing explosive equivalent to about 50 kg TNT) is detonated at a range of 20 m from the opening.

It is possible to undertake an approximate analysis of internal pressure-time histories by making some simplifying assumptions. It is often reasonable to assume that the pressure pulses of both incident and reflected waves are triangular. Therefore, for reflected blast waves

$$p_r(t) = p_r \left(1 - \frac{t}{T_r} \right) \quad 0 \le t \le T_r$$

$$p_r(t) = 0 \qquad\qquad\qquad t \ge T_r \tag{4.11}$$

where T_r is the equivalent positive phase duration of the reflected wave. It should be noted that the pulse duration here is not the same as the actual blast wave positive phase duration T_s from blast parameter graphs. The area under the pressure-time curve for the actual pulse is the specific impulse i_r

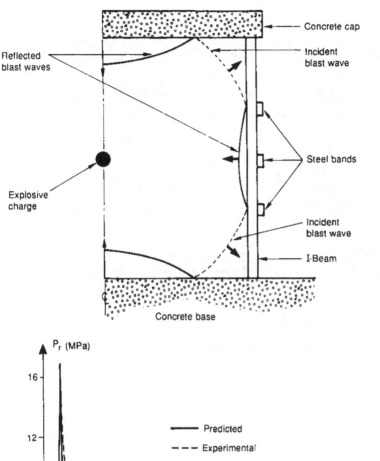

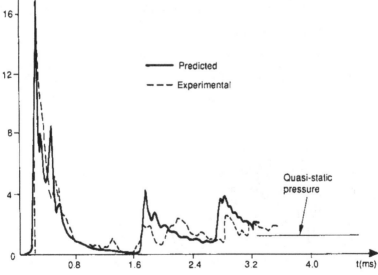

Figure 4.4 *Comparison of pressure-time histories from experiment and analysis for condensed high explosive charge detonated inside a containment structure (after Ref. 11)*

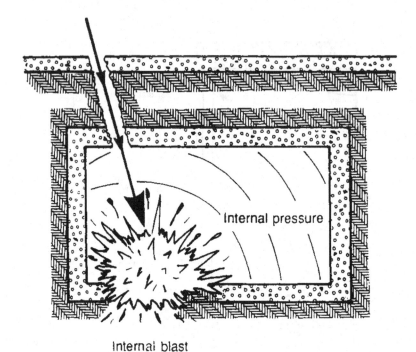

Internal blast

Figure 4.5 *Blast waves generated inside a structure by penetration of a shell or mortar*

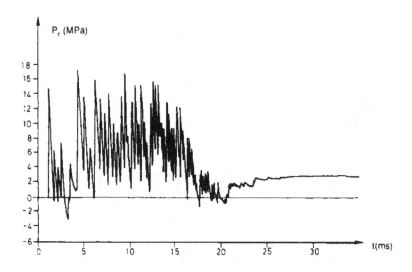

Figure 4.6 *Typical output from BLASTIN*

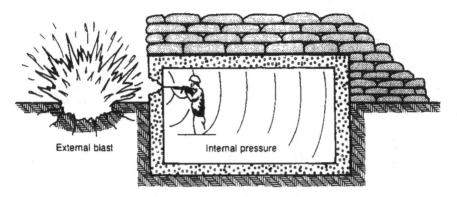

Figure 4.7 *Blast waves entering a structure through an aperture in the structure*

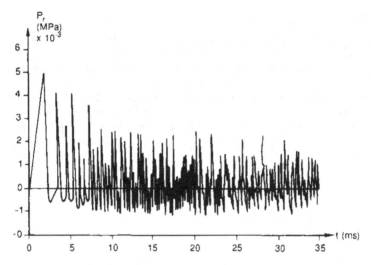

Figure 4.8 *Typical output from CHAMBER*

and this is set equal to the area beneath equivalent triangular pulse. Thus, if the actual reflected specific impulse is i_r then

$$i_r = \tfrac{1}{2}T_r p_r \tag{4.12}$$

and

$$T_r = \frac{2i_r}{p_r} \tag{4.13}$$

Initial internal blast loading parameters are taken to be the normally reflected parameters. This is a reasonable assumption for box-like structures with length to width and height to width ratios of about unity since shock reflections will be regular (rather than Mach) almost everywhere.

Following reflection, re-reflection occurs. A reasonable approach suggested by Baker *et al.*[14] is to assume that the peak pressure is halved on each reflection. Hence the impulse is also halved if duration of each pulse is considered to remain constant. After three reflections, the pressure of reflected wave is assumed to be zero. With reference to Figure 4.9, the situation can be described thus

$$p_{r_2} = \tfrac{1}{2}p_{r_1} \quad p_{r_3} = \tfrac{1}{2}p_{r_2} = \tfrac{1}{4}p_{r_1} \quad p_{r_4} = 0$$
$$i_{r_2} = \tfrac{1}{2}i_{r_1} \quad i_{r_3} = \tfrac{1}{2}i_{r_2} = \tfrac{1}{4}i_{r_1} \quad i_{r_4} = 0 \tag{4.14}$$

where it is assumed that

$$T_r = T_{r_1} = T_{r_2} = T_{r_3} \tag{4.15}$$

In Figure 4.9 the reverberation time – the time delay between each blast wave arriving at the structure internal surface – is assumed constant at t_r (= $2t_a$ where t_a is arrival time of the first blast wave at the reflecting surface). This assumption is not strictly true because successive shocks will be weaker and will thus travel slower than the first.

Further simplification also suggested in Reference 14 can be made particularly if the response time of the structure is a lot longer than the total load duration ($5t_a + T_r$) (see Chapter 8) when all three pulses may be combined into a single pulse having 'total' peak pressure p_{rT} delivering a total specific impulse i_{rT} where

$$p_{rT} = p_{r_1} + p_{r_2} + p_{r_3} = 1.75p_{r_1}$$

$$i_{rT} = i_{r_1} + i_{r_2} + i_{r_3} = 1.75i_{r_1} \tag{4.16}$$

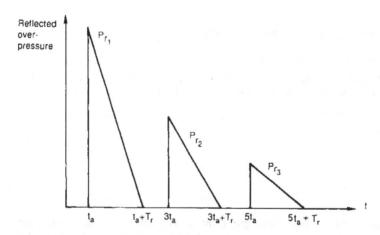

Figure 4.9 *Simplification of blast waves reverberation pressure-time history (after Ref. 14)*

These approximations can be justified in that, when assessing the response of a structure, the use of the approximate input will lead to an overestimate of response leading to a conservative design.

An example of an unvented containment structure capable of withstanding the loading from a 5 kg charge of TNT is given in Figure 4.10 where the massive, heavily reinforced form of the structure should be noted.

While the reverberating blast waves are decaying, the quasi-static or gas pressure is developing. Its magnitude at a particular time will depend on the volume of the structure, the area of any vent in the structure and the characteristics of the particular explosive. A typical pressure-time history for a suppressive structure (a structure with some form of venting) is shown in Figure 4.11.

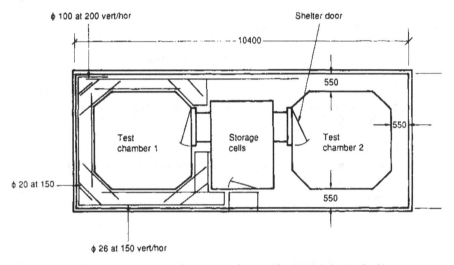

Figure 4.10 *Structure to contain detonation of up to 5kg TNT (after Ref. 12)*

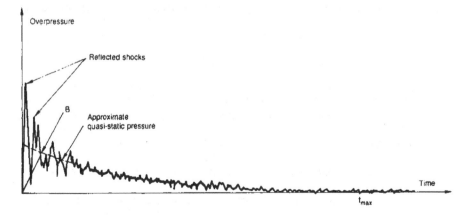

Figure 4.11 *Typical pressure-time history for a vented structure (after Ref. 6)*

The figure shows a series of reverberating blast waves (approximately three in number, confirming the validity of the approach of Equation 4.14) and a developing gas pressure load which peaks at point B and then decays. The precise time when the pressure is reduced to ambient (t_{max} on the graph) is difficult to determine. Reference 14 presents an approach allowing quantification of the important features of the history by use of a simplified form of the gas pressure component of the record as shown in Figure 4.12.

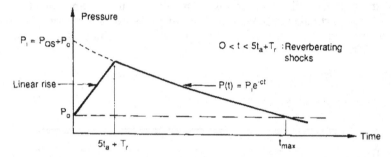

Figure 4.12   *Simplification of development of gas pressure phase inside a structure (after Ref. 14)*

An approximate equation describing the pressure-time history is

$$\bar{p}(t) = \bar{p}_1 e^{(-2.13\tau)} \tag{4.17}$$

where

$$\bar{p} = \frac{p(t)}{p_0} \quad \bar{p}_1 = \frac{p_{QS} + p_0}{p_0} \tag{4.18}$$

in which p_{QS} is the peak quasi-static pressure and p_0 is ambient pressure and

$$\tau = \overline{At} = \left[\frac{\alpha_e A_s}{V^{2/3}}\right]\left[\frac{ta_0}{V^{1/3}}\right] = \frac{\alpha_e A_s t a_0}{V} \tag{4.19}$$

where α_e is the ratio of vent area to wall area, A_s is the total inside surface area of the structure, V is the structure volume and a_0 is the speed of sound at ambient conditions. This equation is valid for the part of the history showing decaying pressure. The rise of gas pressure is assumed to be linear and peaks at a time corresponding the end of the reverberation phase ($5t_a + T_r$). Thus the gas pressure history is the solid line in the figure.

It is worth noting that there are several possible means of venting including permanent openings in the structure, frangible panels that fail at a predetermined overpressure and structures that are designed to fail in such a way as to provide safe venting of an explosion. Figure 4.13 shows such a structure which is designed to fail safe.

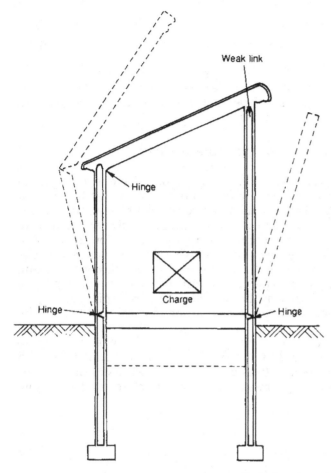

Figure 4.13 *Example of a structure designed to fail safe and vent harmlessly (after Ref. 12)*

From Equation 4.17 it can be shown that the gas pressure falls to ambient pressure at t_{max} corresponding to τ_{max} given by

$$\bar{\tau}_{max} = \frac{1}{2.13} \ln \bar{p}_1 = 0.4695 \ln \bar{p}_1 \qquad (4.20)$$

The area under the curve (ignoring the initial linear rise) is termed the gas impulse i_g which can be written

$$i_g = \int_0^{t_{max}} (p(t) - p_o)dt = \int_0^{t_{max}} (p_1 e^{-Ct} - p_o)dt$$

$$= \left[-\frac{p_1}{C} e^{-Ct} - p_o t \right]_0^{t_{max}} = \frac{p_1}{C}\left[1 - e^{-Ct_{max}} \right] - p_o t_{max} \qquad (4.21)$$

in which

$$C = \frac{2.13\alpha_e A_s a_o}{V} \qquad (4.22)$$

From experimental data from several sources (for example Ref. 6) the curves of Figure 4.14 adapted from References 4 and 5 have been shown to give reasonable predictions of peak quasi-static pressure, 'blowdown' time (t_{max}) and gas pressure impulse.

4.5.2   Internal loading by dust and gas explosions

The pressure pulses generated by the explosion of a fine dust or a gas are characterised by a finite rise time and a duration somewhat longer than would be produced by the detonation of a condensed high explosive. This point is illustrated in Figure 4.15 where pressure-time histories for a 1 kg charge of TNT and a vented confined natural gas explosion in a cubical enclosure containing energy equivalent to 1 kg TNT are compared. A significant reason for the differences lies in the nature of the reactions occurring to produce the overpressure. In the case of dust and gas events the process is one of deflagration, which may be described as very rapid burning, as compared with the detonation process occurring in condensed high explosive material.

Consider firstly the occurrence of a gas explosion. The effects are potentially disastrous as evidenced by the explosion at the Ronan Point flats in London which claimed several lives in 1968. Two distinct situations can be defined. For an explosion to be confined, the structure must be strong

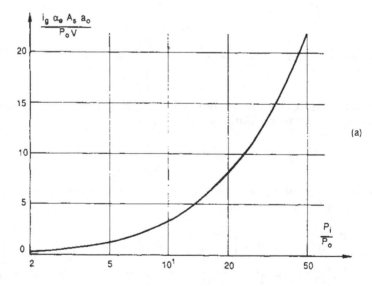

Figure 4.14   (a) *Gas pressure impulse vs normalised maximum pressure*

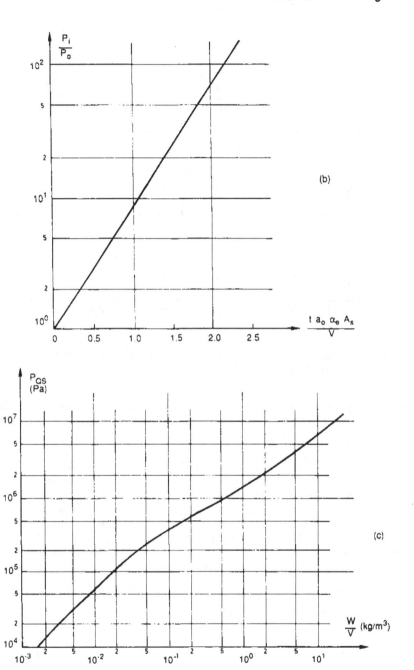

Figure 4.14 (Continued) (b) scaled blow-down duration vs scaled maximum pressure; (c) Peak quasi-static pressure vs charge to volume ratio (after Ref. 14)

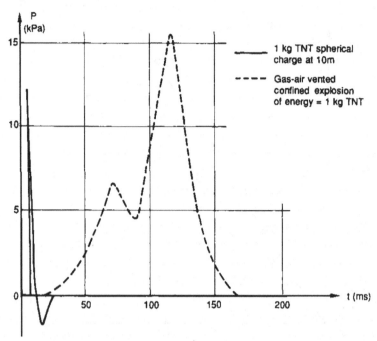

Figure 4.15   *Comparison of pressure time histories for TNT and gas-air mixture of equivalent energy (after Ref. 1)*

enough to prevent venting of explosion products while in a vented-confined explosion the structure contains deliberately weakened elements which allow safe pressure relief.

In a confined explosion, where initially the gas-air mix fills the structure, the maximum pressure p_{max} is a function of the calorific value of the gas. However, it is worth noting that, since stoichiometric gas-air mixes contain approximately the same energy (about 3.6 MJ/m³) no matter what gas is involved in such a mix, p_{max} will be nearly the same at approximately 8 to 9 bars. Because of the finite rise time to p_{max}, the violence of a gas explosion is also measured in terms of the maximum pressure rise rate $(dp/dt)_{max}$. However, this value does differ for different gases because each has a different burn velocity and flame speeds as shown in Table 4.2 adapted from Reference. 1.

In this table MW is molecular weight, $A\%$ is the volume per cent of gas at the lower flammability limit, $B\%$ is the volume per cent of gas producing the maximum burning velocity and CV is the calorific value of the gas at normal temperature and pressure.

A typical pressure-time history for a confined explosion is described by the equation

$$p(t) = p_0 \exp\left\{ E^2(E-1)\left[\frac{S_0 t}{R}\right]^3 \right\}$$

$$(4.23)$$

Table 4.2

Fuel	MW	A%	B%	S_o m/s	T_f K	E	S_f m/s	CV MJ/m^3
Hydrogen	2	30	54	3.5	2318	8.0	28.0	10.2
Methane	16	9.5	10	0.45	2148	7.4	3.5	34
Ethane	30	5.6	6.3	0.53	2168	7.5	4.0	60.5
Propane	44	4.0	4.5	0.52	2198	7.6	4.0	86.4
Butane	58	3.1	3.5	0.50	2168	7.5	3.7	112.4
Hexane	86	2.2	2.5	0.52	2221	7.7	4.0	164.4
Acetylene	26	7.7	9.3	1.58	2598	9.0	14.2	51
Ethylene	28	6.5	7.4	0.83	2248	7.8	6.5	56
Propylene	42	4.4	5.0	0.66	2208	7.7	5.1	81.5
Benzene	56	2.7	3.3	0.62	2287	7.9	4.9	134
Cyclohexane	84	2.3	2.7	0.52	2232	7.8	4.1	167.3

where E is expansion ratio and S_o is burning velocity as defined in Section 4.1, R is the radius of an equivalent sphere of the same volume as the real containment structure and p_o is the initial pressure of the mixture. As is also the case for dusts (see below), $(dp/dt)_{max}$ depends on containment volume as given by

$$\left[\frac{dp}{dt}\right]_{max} = K_G V^{-1/3} \tag{4.24}$$

where V is the volume of the containment structure and K_G is an empirical constant.

In the case of a vented-confined explosion the pressure-time history is characterised by three phases. Initially, the situation is as in a confined explosion. Then any vent cover is removed and finally the venting of the burnt gases occurs. These processes are illustrated in Figure 4.16.

Consider now the occurrence of dust explosions which are described by Bartknecht[7] and summarised by Goschy[8] and Kinney and Graham.[9] Wherever there is fine particulate material and an initiating source in proximity such an event is possible. Flour mills, coal mines, chemical and pharmaceutical plants represent likely locations. For example, there are between 30 and 40 grain elevator explosions in the United States each year. Two of the worst cases occurred in Westwego, Louisiana, and Galveston, Texas (both in 1977), when 36 and 18 people respectively were killed.

The sequence of events leading to a dust explosion could involve a small electrical explosion in a piece of equipment in, say, a flour mill. The equipment may rupture and sparks may be generated setting dust on or near the equipment alight. The flame thus produced could spread and more dust could be stirred up, leading to a large-scale explosion which could have devastating consequences.

Dust particles involved in an explosion are typically of diameter 75 μm or less. The concentration of dust needed to cause an explosion is generally high: visibility through the dust cloud may be as little as 0.2 m. This

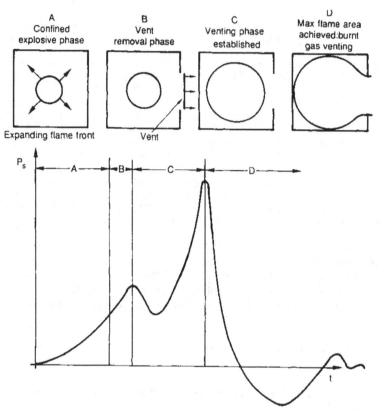

Figure 4.16   *Course of events in a gas explosion in a vented structure (after Ref. 1)*

means there is an almost opaque atmosphere which is not usually found in a working area where personnel are present. The lower limit for an explosion to occur is a concentration of between 20 and 60 g/m^3 while the upper limit is somewhere between 2 and 6 kg/m^3.

The pressure-time history of such an event is usually described as for a gas explosion in terms of the 'violence' measured by the maximum rate of pressure rise $(dp/dt)_{max}$, as well as the maximum pressure generated p_{max}, both quantities being functions of concentration. Figure 4.17 shows how rise rate is defined, where it is worth noting that greatest explosive violence occurs at concentrations in excess of the stoichiometric mix of dust and air: usually charred remains are evident after a very violent explosion indicating an excess of fuel (dust) compared with oxidising agent (air).

Figure 4.18 shows how maximum pressure and pressure rise rate vary with dust concentration.

Other factors that affect violence include particle size (characterised either by particle diameter or particle surface area), the initial pressure of the dust-air mix and the size of the structure in which the explosion occurs. With regard to particle size it is found that fine dusts are more violently explosive than coarse dusts. The greatest pressures are produced with the characteristic

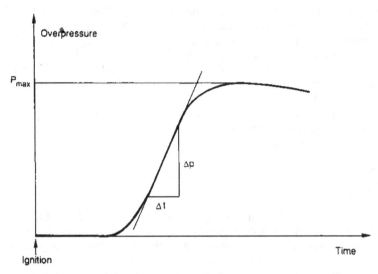

Figure 4.17 *Definition of maximum pressure rise rate in gas and dust explosions (after Ref. 8)*

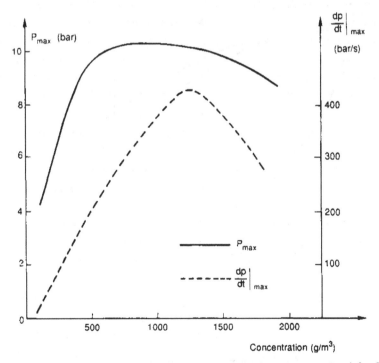

Figure 4.18 *Maximum pressure and pressure rise rate vs dust concentration (after Ref. 7)*

particle size less than 40 μm. If the size (usually the d_{50} value, the size of sieve passing 50% of dust particles in a sample) exceeds 400 μm there is little chance of an explosion occurring. Particle surface area rather than d_{50} may be a better measure of size because a larger particle surface area means that more oxygen is in contact with the fuel. These points are illustrated in Figures 4.19 and 4.20. If the initial pressure of the dust-air mix is high, violence will increase as shown in Figure 4.21.

Finally, the influence of the size of the confining structure is represented by an empirical equation analogous to Equation 4.24, which links maximum pressure rise rate and confining volume V

$$\left[\frac{dp}{dt}\right]_{max} = K_D V^{-1/3} \tag{4.25}$$

where K_D is a material constant. Values for some dusts are given in Table 4.3 adapted from Reference 8 which also gives approximate maximum pressures.

It should be noted that, in determining the most violent explosion by combustible dusts, a minimum volume of containment structure is needed in order to develop full reaction rate as will be seen from Figure 4.22. For this reason, the standard laboratory test vessel is of approximately 16 litre capacity.

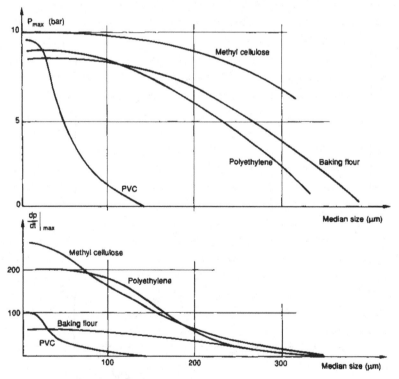

Figure 4.19 *Maximum pressure and pressure rise rate vs particle size (after Ref. 7)*

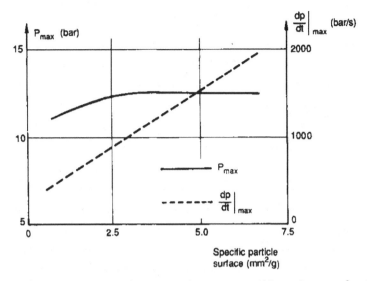

Figure 4.20 *Maximum pressure and pressure rise rate vs particle surface area for aluminium dust (after Ref. 7)*

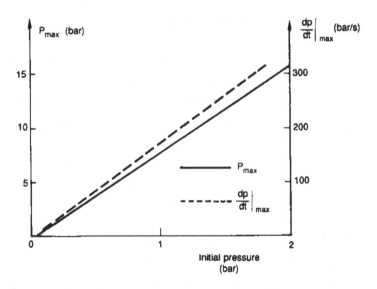

Figure 4.21 *Maximum pressure and pressure rise rate vs initial pressure for starch dust (after Ref. 7)*

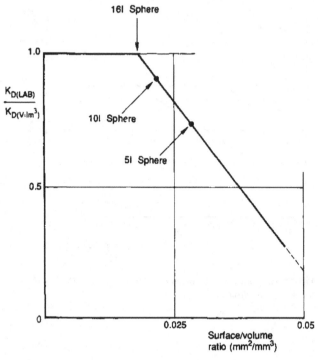

Figure 4.22 *Development of a dust explosion: dependency on containing vessel volume (after Ref. 7)*

Table 4.3

Dust	p_{max} (bar)	K_D(bar/m/s)
PVC	~8	27–98
Milk powder	~9	58–130
Sugar	~9	59–165
Coal	~9+	93–176
Pigments	6–11	28–334

4.6 References

[1] Harris R.J., *Gas Explosions in Buildings and Heating Plant*. E&FN Spon, London (1983)
[2] Institution of Chemical Engineers Major Hazards Assesment Panel (Overpressure Working Party). *The Effects of Explosions in the Process Industries*. Institution of Chemical Engineers (1989)
[3] Baker W.E., Hu W.C.L., Jackson T.R., Elastic Response of Thin Spherical Shells to Axisymmetric Blast Loading. *Journal of Applied Mechanics*, Trans ASME, pp. 80–6, December (1966)

[4] Britt J.R., Drake J.L., Cobb M.B., Mobley J.P., BLASTINW User's Manual ARA 5986-2 Contract DACA39-86-M-0213 for USAE Waterways Experiment Station, Applied Research Associates Inc. (April 1986)

[5] Britt J.R., Drake J.L., Cobb M.B., Mobley J.P., CHAMBERM User's Manual ARA 5986-1 Contract DACA39-86-M-0213 for USAE Waterways Experiment Station, Applied Research Associates Inc. (April 1986)

[6] Weibull H.R.W., Pressures recorded in partially closed chambers at explosion of TNT charges. *Annals of the New York Academy of Sciences*, Vol. 152, Art. 1, pp. 357–61 (1968)

[7] Bartknecht W., *Explosions: Cause, Prevention and Protection*. Springer-Verlag, Berlin (2nd edition) (1982)

[8] Goschy B., *Design of Buildings to Withstand Abnormal Loading*. Butterworth-Heinemann, Oxford (1990)

[9] Kinney G.F., Graham K.J., *Explosive Shocks in Air*. (2nd edition) Springer-Verlag, New York (1985)

[10] Axelsson H., Berglund S., Cloud development and blast wave measurements from detonation Fuel Air Explosive charges. National Defence Research Institute, Stockholm Report FAO C 20225-D4 (1978)

[11] Gregory F.H., Analysis of the loading and response of a suppressive shield when subjected to an internal explosion. Minutes of the 17th Explosives Safety Seminar, Denver, Colorado (1976)

[12] Bergman S.G.A., Swedish protective structures for manufacturing units constituting explosion hazard in the range 1–2000lbs TNT. *Annals of the New York Academy of Sciences*, Vol. 152, Art. 1, pp. 500–9 (1968)

[13] BL1-5 Blast loading: Joint industry project on blast and fire engineering for topside structures. The Steel Construction Institute (1992)

[14] Baker W.E., Cox P.A., Westine P.S., Kulesz J.J., Strehlow R.A., *Explosion Hazards and Evaluation*. Elsevier, Amsterdam (1983)

[15] TM5-855-1. Fundamentals of Protective Design for Conventional Weapons. *US Department of the Army Technical Manual* (1987)

Symbols

a_o	speed of sound at ambient conditions
A_f	flame area
A_n	minimum flame area
A_s	total inside surface area of structure
E	expansion ratio
i_g	specific gas impulse
i_r	specific reflected overpressure impulse
K_D	empirical constant for violence of dust explosions
K_G	empirical constant for violence of gas explosions
m_b	mass of burnt gas
N_b	moles of burnt gas
N_u	moles of unburnt gas
$\bar{p}$	scaled pressure
p_o	atmospheric pressure
p_{QS}	quasi-static pressure
p_r	peak reflected overpressure
p_s	peak side-on overpressure

$(dp/dt)_{\text{max}}$	maximum rate of pressure rise
r	radial distance
R	radius of equivalent sphere
S_f	laminar flame speed
S_o	laminar burn velocity
S_T	turbulent burn velocity
t	time
t_a	arrival time of blast wave
t_{max}	blowdown time
T_f	flame temperature
T_i	initial gas temperature
T_r	positive phase duration of reflected overpressure
V	structure volume
V_b	volume of burnt gas
V_u	volume of unburnt gas
α_e	vent area ratio
β	empirical turbulence factor
ρ_b	density of burnt gas
ρ_u	density of unburnt gas
τ	scaled time

5 Underwater explosions

5.1 Introduction

The detonation of a condensed high explosive underwater produces two pressure pulses: a shock wave followed by a bubble pulse associated with the expansion of the products of detonation. For the shock wave, analysis proceeds along the same lines as for airblast although there are quantitative differences in the results. For the bubble pulse, the sequence of events is described below.

The bubble (comprising hot, compressed gases) expands, pushing water away from the source of the explosion. The bubble radius increases and 'overshoots' beyond the equilibrium radius corresponding to the hydrostatic pressure at the current depth of the bubble. This is because of the effect of the inertia of the water set in motion by the pulse. Then, as this overshoot occurs, the pressure in the bubble becomes less than local hydrostatic pressure. This situation does not persist and the bubble collapses due to the excess hydrostatic pressure now being exerted and the gas is recompressed, now 'undershooting' to a radius smaller than that corresponding to local conditions and the bubble pressure exceeds local hydrostatic pressure leading to expansion. Any subsequent expansions and contractions are damped out as energy is dissipated and the bubble rises to the surface.

This train of events is illustrated in Figure 5.1 which shows both the shock wave pulse and the variation of the bubble size with time and depth. Generally the positive phase duration of the bubble τ_b is greater than that of the shock wave T^+.

5.2 Underwater shock-wave details

If the maximum pressure at the shock wavefront is written p_m, the time variation of pressure in the wave is usually described by the equation

$$p(t) = p_m e^{-\frac{t}{\theta}} \tag{5.1}$$

where the time constant θ is time for $p(t)$ to fall to p_m/e. The impulse delivered by the shock I is given by

$$I = \int_0^{5\theta} p(t)dt \tag{5.2}$$

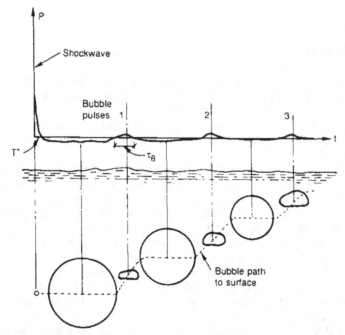

Figure 5.1   *Pressure-time history for underwater explosion showing shock wave and bubble pulse (after Ref. 2)*

The form of $p(t)$ means that the upper limit of integration (here 5θ) must be arbitrarily fixed to allow a realistic value of I to be obtained. Sometimes this limit is taken as 6.7θ.

The energy flux of the shock wave (the 'energy') E is obtained by considering a volume of fluid of cross-sectional area A acted on by the shock wave such that a particle velocity u is imparted to the water. Thus, in an increment of time dt, the distance moved by the fluid is $u \cdot dt$ as shown in Figure 5.2.

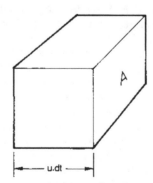

Figure 5.2   *Elemental volume for shock wave analysis*

So, if the pressure acting at time t is $p(t)$, the force on the area A is $p(t)A$ and the work done will be the energy of the fluid element dE given by

$$dE = p(t)Au\,dt \tag{5.3}$$

Thus the total work done, which is the total energy of the fluid E, is given by

$$E = \int_0^{5\theta} p(t)Au\,dt \tag{5.4}$$

Now, assuming that water is an elastic medium, the transmission of the shock wave may be described by an equation analogous to the transmission of a stress wave through a solid elastic medium (Chapter 6) in which the stress σ is given by

$$\sigma = \rho c u \tag{5.5}$$

where ρ is the density of the medium and c is the speed of sound in the medium. In this case the relationship between pressure and particle velocity will be of the form

$$p(t) = \rho_w c_w u \tag{5.6}$$

where ρ_w is the density of water and c_w is the speed of sound in water which is approximately 1400 m/s. Thus

$$u = \frac{p(t)}{\rho_w c_w} \tag{5.7}$$

and the energy flux density (or the energy per unit area) E' is

$$E' = \int_0^{5\theta} \frac{p(t)^2}{\rho_w c_w}\,dt \tag{5.8}$$

or

$$E' = \frac{1}{\rho_w c_w} \int_0^{5\theta} p(t)^2 dt \tag{5.9}$$

The variation of important shock wave parameters can be represented on a nomogram such as that shown in Figure 5.3. As in the case of airblast, charge weight is given in terms of kilogrammes of TNT. For other explosives conversion to TNT is the step prior to using the nomogram.

The use of Figure 5.3 is best illustrated by means of an example.

Example 5.1
A 1000 kg spherical TNT charge is detonated underwater. What are the shock wave parameters at a range of 10 m?

Firstly, draw a straight line from the 1000 kg level on the right hand column to the 10 m mark on the left hand column. From the interceptions of this line with the maximum pressure, impulse and energy columns obtain p_m as 52 MPa, I as approximately 57 kPa-s, and E' as about 800 kPa-m. Secondly, draw a line from 1000 kg to the 10 m mark on the second column from the left and obtain the time constant θ as 0.82 ms.

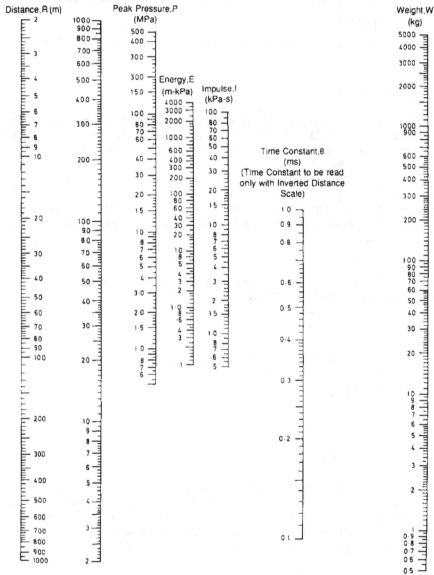

Figure 5.3   *Nomogram for shock wave parameter evaluation (after Ref. 2)*

There are also equations fitted to experimental data for p_m and I (of the same general form as presented for airblast) expressed in terms of scaled distance Z based on range and charge weight of TNT. Examples are given below for p_m.

$$p_m = \frac{355}{Z} + \frac{115}{Z^2} - \frac{2.44}{Z^3} \text{bar} \quad 0.05 \leq Z \leq 10$$

$$p_m = \frac{294}{Z} + \frac{1387}{Z^2} - \frac{1783}{Z^3} \text{bar} \quad 10 \leq Z < 50 \tag{5.10}$$

Finally, graphs for p_m and I plotted against scaled distance which have been derived from a number of experimental sources are shown in Figure 5.4.

As noted above, to use these graphs the explosive must be converted to TNT. It is also possible to deal with explosives other than TNT by using a scaled distance Z' equal to $R/W_e^{1/3}$ where W_e is now the actual explosive mass in kilograms. Thus we can write

$$p_m = KZ'^{-\alpha} \tag{5.11}$$

where K and α are empirical constants for specific explosives. Typical values are given in Table 5.1 (adapted from Reference 2) where K is given to evaluate p_m in megaPascals.

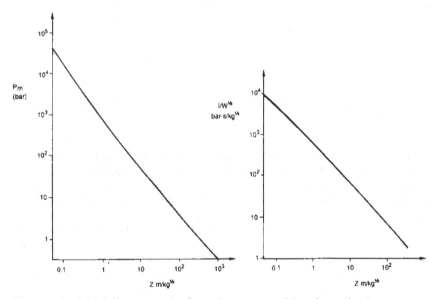

Figure 5.4 *Scaled-distance graphs for peak pressure and impulse evaluation*

Table 5.1

Explosive	K	α	Valid for
TNT	52.4	1.13	3.4 to 138 MPa
Pentolite	56.6	1.14	3.4 to 138 MPa
H-6	59.2	1.19	10.3 to 138 MPa

5.3 Bubble parameters details

The bubble pulse is caused by the expansion of the explosive gases which also rise to the surface as time passes. Figure 5.5 shows the variation of radius with time for a gas bubble formed by the detonation of 250 g tetryl under water at a depth of 90 m.

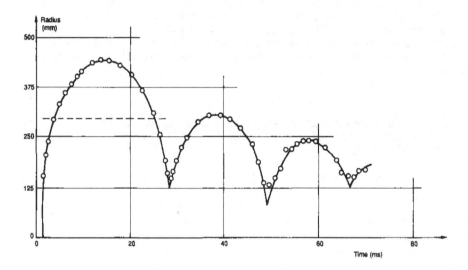

Figure 5.5 *Variation of bubble diameter with time for a 250 g tetryl charge detonated at 90 m depth (after Ref. 2)*

The dotted horizontal line in the figure shows the radius at which internal pressure and outside hydrostatic pressure at the particular bubble location are equal, demonstrating the overshooting and undershooting in radius beyond the 'equilibrium' radius as the bubble expands and contracts.

The analysis of the bubble motion (adapted here from Reference 1) assumes firstly that water is an incompressible medium and that secondly the water motion is entirely radial. Using the same approach as in the analysis of airblast, solution of the equations of mass and momentum conservation together with appropriate boundary conditions is required. Thus, from airblast analysis recall that the mass conservation equation for a small element (Equation 3.2) is written

$$\frac{\partial \rho}{\partial t} + u \frac{\partial \rho}{\partial r} + \rho \frac{\partial u}{\partial r} + \frac{\rho u}{A} \frac{\partial A}{\partial r} = 0 \tag{5.12}$$

For spherical bubbles the appropriate surface area A is $4\pi r^2$ where r is radial distance from which

$$\frac{\partial A}{\partial r} = 8\pi r \tag{5.13}$$

and hence

$$\frac{\partial \rho}{\partial t} + u \frac{\partial \rho}{\partial r} + \rho \frac{\partial u}{\partial r} + \frac{2\rho u}{r} = 0 \tag{5.14}$$

The conservation of momentum equation (Equation 3.7) is written

$$\frac{\partial u}{\partial t} + u \frac{\partial u}{\partial r} + \frac{1}{\rho} \frac{\partial p}{\partial r} = 0 \tag{5.15}$$

If water is assumed incompressible in this analysis, ρ is constant and Equation 5.14 becomes

$$\frac{\partial u}{\partial r} = -\frac{2u}{r} \tag{5.16}$$

and so we can write

$$u(r, t) = \frac{u_1}{r^2} \tag{5.17}$$

where u_1 is a constant and is the velocity when r is unity. In equation 5.15 therefore, on substitution

$$\frac{1}{r^2} \frac{du_1}{dt} - \frac{2u^2}{r} + \frac{1}{\rho} \frac{\partial p}{\partial r} = 0 \tag{5.18}$$

which can be rewritten using Equation 5.16 as

$$\frac{1}{r^2} \frac{du_1}{dt} + \frac{1}{2} \frac{\partial(u^2)}{dr} + \frac{1}{\rho} \frac{\partial p}{\partial r} = 0 \tag{5.19}$$

We now wish to integrate from the gas sphere surface where

$$r = a, u_a = \frac{da}{dt} = \frac{u_1}{a^2} \quad \text{and} \quad p = p_a \tag{5.20}$$

to infinity where pressure is ambient pressure p_o and velocity is zero. Thus

$$\int_a^\infty \frac{1}{r^2} \frac{du_1}{dt} dr + \int_a^\infty \frac{1}{2} \frac{\partial u^2}{\partial r} dr + \frac{1}{\rho} \int_a^\infty \frac{\partial p}{\partial r} dr = 0 \tag{5.21}$$

or, with appropriate substitution

$$\int_a^\infty \frac{1}{r^2} \frac{d}{dt} \left[a^2 \frac{da}{dt} \right] dr + \int_a^\infty \frac{1}{2} \frac{\partial u^2}{\partial r} dr + \frac{1}{\rho} \int_a^\infty \frac{\partial p}{\partial r} dr = 0 \tag{5.22}$$

which can then be rewritten as

$$\frac{1}{a} \frac{d}{dt} \left[a^2 \frac{da}{dt} \right] - \frac{1}{2} \left[\frac{da}{dt} \right]^2 - \frac{1}{\rho} (p_a - p_o) = 0 \tag{5.23}$$

The equation is now multiplied through by ρ to give

$$\rho \frac{1}{a} \left[a^2 \frac{da}{dt} \right] - \frac{\rho}{2} \int_0^t \left[\frac{da}{dt} \right]^2 dt - \int_0^t p_a dt + \int_0^t p_o dt = C \tag{5.24}$$

and is in a form to allow integration with respect to time, term by term. After integration and simplification as detailed in References 1 and 2, the final result is

$$\frac{1}{2}\rho a^3\left[\frac{da}{dt}\right]^2 + \frac{1}{3}p_0 a^3 - \int_0^a p_a a^2 da = C'$$

(5.25)

The last term on the left hand side is the work done in expanding the bubble against hydrostatic pressure. This term can be thought of as representing a decrease in the internal gas energy to a value $E(a)$. Now, noting that gas bubble volume V is

$$V = \frac{4}{3}\pi a^3 \qquad \frac{dV}{da} = 4\pi a^2$$

$$\therefore \quad dV = 4\pi a^2 da$$

(5.26)

the final term can be written

$$\frac{1}{4\pi}\int_0^a p_a dV = \frac{1}{4\pi}[E(a) - E(0)]$$

(5.27)

So if Equation 5.25 is rearranged and the initial internal gas energy value is absorbed into the constant of integration we obtain

$$\frac{3}{2}\left[\frac{4\pi}{3}\rho a^3\right]\left[\frac{da}{dt}\right]^2 + \frac{4}{3}\pi p_0 a^3 + E(a) = \Upsilon$$

(5.28)

where Υ can be thought of as the total bubble energy.

Now consider the lefthand side of this equation term by term. The first, which has the form ($\frac{1}{2}$ × mass × velocity^2) represents the kinetic energy of the bubble. The second represents the work done against hydrostatic pressure and the third is the internal energy of the bubble after significant expansion has occurred. This quantity can be shown as proportional to $(1/a^3)$ and decreases rapidly with bubble radius and may be ignored.

A most important parameter is the maximum bubble radius a_{max}, occurring when the bubble expansion rate, da/dt, is zero. Therefore, in equation 5.28 with both $E(a)$ and da/dt set to zero the total bubble energy may be written

$$\Upsilon = \frac{4}{3}\pi p_0 a_{max}^3$$

(5.29)

Now, using Equations 5.28 and 5.29 to eliminate Υ, numerical integration of the resulting equation will allow the half period of oscillation (the time to reach a_{max}), T' and a_{max} to be evaluated. The equation to be solved is

$$\left(\frac{da}{dt}\right)^2 = \frac{2p_0}{3\rho}\left(\left(\frac{a_{max}}{a}\right)^3 - 1\right)$$

(5.30)

When the variables are separated the equation becomes

$$\int_0^{a_{max}} \frac{da}{\sqrt{\left(\frac{a_{max}}{a}\right)^3 - 1}} = \sqrt{\frac{2p_0}{3\rho}} \int_0^{T'} dt \tag{5.31}$$

and integration leads to

$$2T' = \tau_b = \frac{1.14\rho^{1/2}\Upsilon^{1/3}}{p_0^{5/6}} \simeq \frac{K_{ex}W^{1/3}}{(H + H_o)^{5/6}}$$

$$a_{max} = \frac{J_{ex}W^{1/3}}{(H + H_o)^{1/3}} \tag{5.32}$$

Because total bubble energy can be related to the weight of charge being detonated, a convenient form for τ_b and a_{max} is given in Equation 5.32 where K_{ex} and J_{ex} are empirical constants that depend on the particular explosive. For TNT K_{ex} is 2.11 $sm^{5/6}$ $kg^{-1/3}$ and J_{ex} is 3.50 $m^{4/3}$ $kg^{-1/3}$. In Equation 5.32, H is charge depth in metres, H_o is atmospheric head (approximately 10 m of water) and W is charge weight in kilogrammes. Figure 5.6 is a nomogram for estimation of τ_b and a_{max}.

5.4 Partition of energy

As was mentioned above both shock wave and bubble pulse components of an underwater explosive event contain significant amounts of the total energy released. Figure 5.7 is a schematic breakdown of how energy is partitioned between the two components.

Both pressure components are important in structural loading as will be seen in Table 5.2 (adapted from Reference 2) which shows a comparison of the shock wave and bubble pulse pressures, impulses and energies measured at 20 m from 135 kg TNT in 30 m water.

Even though the bubble pulse may not deliver a particularly high pressure to a structure, when coupled with a long duration (recall Figure 5.1 above) the damage to an underwater target structure can be as significant as that caused by the shock component which is likely to have a higher peak over-pressure but will be of much shorter duration.

As a specific example consider TNT for which the shock wave energy has been measured at 2345 kJ/kg and the bubble energy at 2000 kJ/kg. The total energy is thus 4345 kJ/kg which is very close to the heat of detonation of TNT which is approximately 4520 kJ/kg.

5.5 Surface and sea-bed interactions

It is worth noting that two sorts of interaction in relation to the shock wave will occur. When it interacts with the effectively rigid sea bed, enhancement of the overpressure will take place just as for a blast wave in air. However,

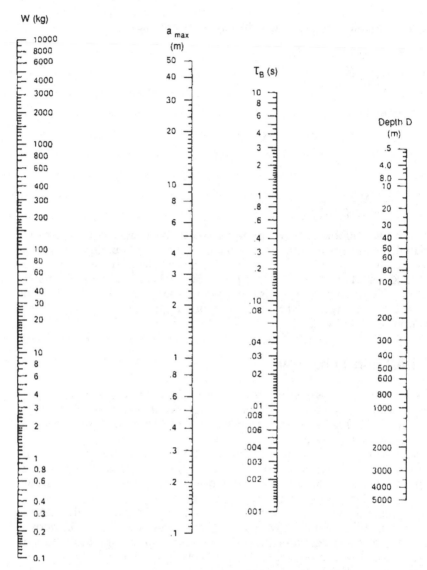

Figure 5.6 *Nomogram for evaluation of bubble pulse parameters (after Ref. 2)*

Table 5.2

	Depth (m)	p_m (kPa)	I (kPa-s)	E (kPa-m)
Shock wave	12.2	12 204	7.93	29.8
Direct (first)	13.7	386	10.34	1.7
Bubble pulse				
Composite	13.7	–	17.93	3.3
Bubble pulse				

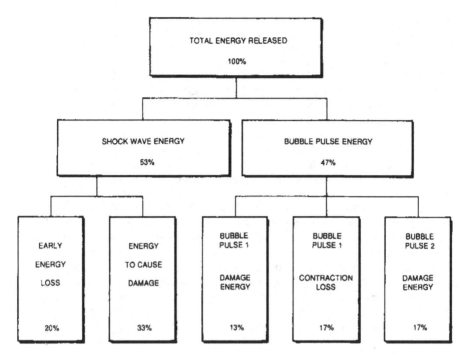

Figure 5.7 *Division of energy from an underwater explosion*

when the shock wave arrives at the surface of the water (the water/air interface) rarefaction occurs and the basic shock wave profile is modified as shown in Figure 5.8.

Hence it is seen that the impulse loading on a near-surface target will be less than that experienced by a submerged target. In other words, submarines are likely to be more severely loaded than surface vessels.

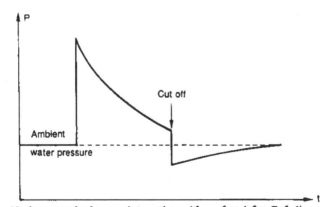

Figure 5.8 *Underwater shock wave interaction with surface (after Ref. 4)*

In the specific case of underwater nuclear explosions two categories are identified. Firstly, in a shallow explosion the detonation is either close to the surface or the depth of water is small, while in a deep explosion detonation is well below the surface or the water is deep.

The sequence of events in a shallow explosion is that, following detonation and the accompanying small fireball, a bubble of hot gases is produced and a shock wave is initiated. Soon after, a mound or column of water (the spray dome) forms and there is partial transmission of the shock wave to the air to form airblast. A condensation cloud then forms above the surface and then the bubble breaks surface and a hollow chimney forms (sometimes called the 'cauliflower' effect). A base surge (a dense water droplet cloud) then develops and a series of large amplitude water surface waves propagates. Conversely, in a deep water explosion following detonation there is little evidence of fireball. A nearly spherical gas bubble forms and the shock wave propagates as the bubble rises and oscillates and in so doing possibly forms new shocks. Some of these features are illustrated in Figure 5.9.

Returning to conventional explosive material, a technique for the protection of submerged structures from the shock wave can make use of the same phenomenon noted above in the use of bubble screens or bubbles curtains.[5] As an example, consider an offshore gas or oil production platform. It is sometimes necessary to repair or modify submerged parts of the structure by using explosive cutting charges. It is clear that explosive

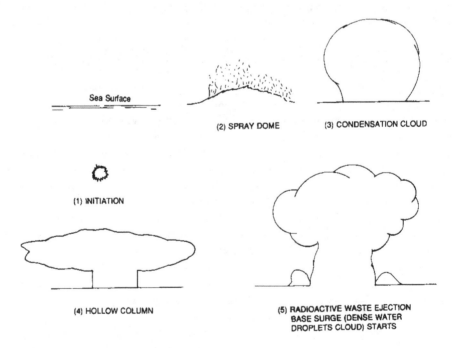

Figure 5.9   *Development of base surge from a deep nuclear explosion*

loading should be confined to the component being cut and any dynamic loading on adjacent structural elements should be kept small. Localisation of the shock wave effect could be achieved by shrouding the work area in a screen of gas bubbles by using, perhaps, gas from a diver's bottle. When the charge is fired the outward moving shock wave interacts with the surrounding bubble barrier. The many reflection surfaces provided by the highly irregular bubble flow leads to considerable attenuation of the shock wave energy.

The problem is not so simple regarding the bubble pulse component. In this case the loading is provided by the rapid movement of a dense fluid. This motion will not be suppressed by a bubble screen. A more substantial barrier would be required to protect the adjacent structure from the potentially damaging flow. These ideas are illustrated in Figure 5.10.

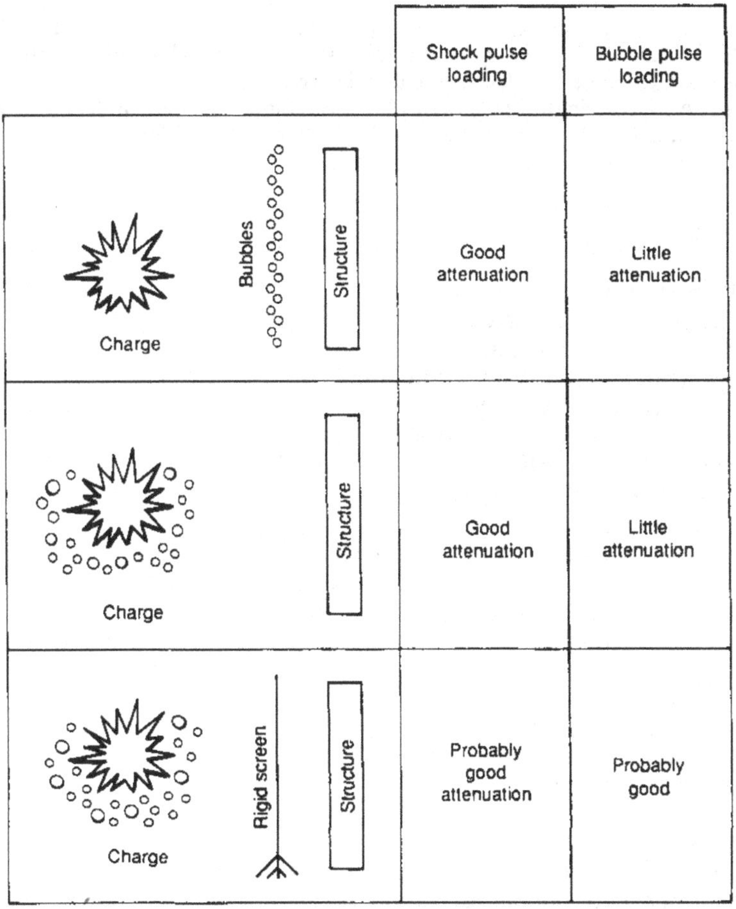

Figure 5.10 *Techniques for reducing shock and bubble pulse effects on targets*

5.6　Further reading

The most well-known text concerned with underwater explosions and their effects is that by Cole.[1] Key elements of this work are presented in the form of a summary which is available as part of Reference 2. The book by Henrych[3] also provides a brief mathematical treatment of the subject. For a more descriptive approach Reference 4 includes photographs from full-scale trials showing features of deep and shallow underwater bursts.

5.7　References

[1] Cole R.H., *Underwater Explosions*. Dover, New York (1948 reprinted 1963)
[2] Kaye S.H., (ed.)., *Encyclopaedia of Explosions and Related Terms*. PATR Vol. 10. United States Research and Development Command Large Caliber Weapons Systems Lab. NJ (1983)
[3] Henrych J., *The Dynamics of Explosion and Its Use*. Elsevier, Amsterdam (1979)
[4] Glasstone S., Dolan P.J., *The Effects of Nuclear Weapons*. United States Department of Defense and United States Department of Energy (1977)
[5] Barnes R.A., Hetherington J.G., Smith P.D., Bubble screens for underwater shock attenuation. *Explosives Engineering*, Vol. 2, No. 3, pp. 6–9 (1988)

Symbols

a	bubble radius
a_{max}	maximum bubble radius
A	fluid cross-sectional area
c	speed of sound in elastic medium
c_w	speed of sound in water
E	energy flux of underwater shock wave
E'	energy flux density of underwater shock wave
$E(a)$	internal energy of gas bubble
H	depth of detonation of charge
H_0	atmospheric head (= 10 m of water)
I	specific impulse delivered by underwater shock wave
J_{ex}	empirical explosive constant
K	empirical constant in maximum pressure-scaled distance equation
K_{ex}	empirical explosive constant
p	pressure
p_m	maximum pressure at shock wave front
p_a	pressure at gas bubble surface
p_0	ambient pressure
r	radial distance
t	time
T^+	underwater shock wave positive phase duration
T'	half period of bubble oscillation
u	particle velocity

u_1 particle velocity when $r = 1$

V gas bubble volume

W charge mass (expressed in kg TNT)

W_e actual charge mass

Z scaled distance

Z' scaled distance based on actual charge mass

α empirical constant in maximum pressure-scaled distance equation

θ time constant

ρ density

ρ_w density of water

σ stress

τ_b bubble pulse positive phase duration

Υ total bubble energy

6 Stress waves

6.1 Introduction

We have already seen the way in which airborne blast can damage structures. A vertical wall in the path of a blast wave will experience a loading derived from both the static overpressure and the dynamic overpressure due to the concerted movement of the air molecules impinging on the structure. These pressures will cause the wall to bend and deflect and damage to the structure will result if the stresses or strains which ensue exceed those which the material can tolerate. At the same time an associated phenomenon is taking place. The stress level on the front face of the wall rises and falls suddenly due to the arrival of the airborne pressure wave. This transient stress is propagated through the wall in the form of a compressive *stress wave*. On arrival at the rear face of the wall it is reflected back as a tensile wave. This process is described schematically in Figure 6.1, with compressive stresses being taken as positive and tensile stresses as negative. During the reflection process, the leading edge of the tensile reflected wave overlaps the trailing edge of the incoming compressive wave. The resultant stress experienced by the material, at a specific location, is therefore the algebraic sum of the two waves at that point at that instant in time. The intensity of the

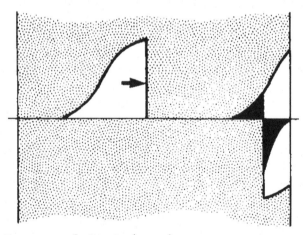

Figure 6.1 *Stress wave reflection at a free surface*

nett tensile stress therefore builds as the tensile wavefront proceeds back towards the face of the target until it achieves its full value when it passes the tail of the incoming compressive wave. Nearly all materials are weaker in tension than compression and thus a wave, which has passed uneventfully through the target, may well tear a scab from the rear face. Experience shows us that for an airborne blast wave from a source remote from the wall, scabbing is unlikely to occur. However, if the source of the explosion is near to, or in contact with, the slab, scabbing becomes the more likely form of failure. Scabbing can also occur in both structures and armoured fighting vehicles as a result of remotely delivered explosive charges, known as high explosive squash head rounds (HESH), and indeed from ballistic impact by inert projectiles.

A simple one-dimensional wave model will provide us with a sufficient insight into the behaviour of stress waves to enable us to design structures which are resistant to this form of attack. Our study of groundborne stress waves, or *groundshock*, will include a more comprehensive model, covering compressive, shear and surface waves (Chapter 7).

6.2 Stress wave parameters

In describing the propagation of a stress wave, we need to recognise that, within the system, stress varies with both time and position. Thus if we take a 'snapshot' view of the stress distribution through the thickness of a plate, at a particular instant in time, we may obtain, for example, a variation of the form shown in Figure 6.2a.

However, as an observer recording the variation of stress with time at a particular location, one would find a variation of the form shown in Figure 6.2b. In practice it is easier to analyse rectangular pulses and we therefore break down the real pulse into a series of pulses as shown in Figure 6.3a. A further simplification is achieved by representing a rectangular pulse as a sequence of two wavefronts (or step changes in stress), of equal magnitude

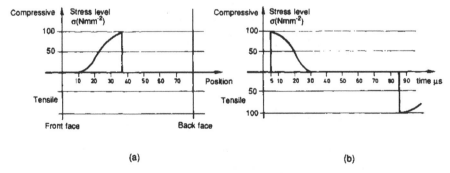

(a)   (b)

Figure 6.2 *(a) Variation of stress with position at a particular instant; (b) variation of stress with time at a particular location*

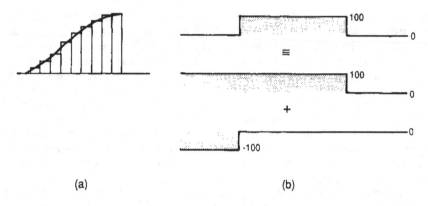

(a)   (b)

Figure 6.3   *(a) Representation of a pulse by a series of rectangular pulses; (b) representation of a rectangular pulse*

but opposite sign, separated by a time interval equal to the duration of the pulse (Figure 6.3b). The first wavefront establishes the stress, the second removes it. By these two simple devices, we may infer the behaviour of a real pulse from an understanding of the behaviour of individual wavefronts. Figures 6.4a and b represent a simplified model of a solid material. The trucks represent the molecules of the material and the springs the forces which hold them together and apart at their equilibrium spacing. When a force F is suddenly applied and sustained the force pushes on the first truck which pushes on the second truck and so on down the line. The message is passed rapidly from truck to truck and the speed at which the message travels down the line is called the *wave propagation velocity, c*. Each truck in turn starts to move, but the velocity at which the trucks move, the *particle velocity, v*, will be significantly less than the propagation velocity. It is interesting to note that, in the case of the compressive wave generated in Figure 6.4a, the particle velocity and the propagation velocity are in the same direction. For the tensile wave of Figure 6.4b, however, the trucks move to

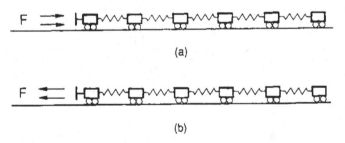

Figure 6.4   *(a) Transmission of a compressive wavefront; (b) transmission of a tensile wavefront*

the left whilst the wave travels to the right – particle and propagation velocities are in opposite directions. Having differentiated between particle and propagation velocities, we will now determine their magnitudes.

6.2.1 Propagation velocity

Figure 6.5 represents an element of a solid bar, of cross-sectional area A and density ρ, through which is propagating a stress wave. At the point x, the stress is σ, but, at a short distance, dx, further down the bar the stress is $\left(\sigma + \frac{\partial \sigma}{\partial x} \cdot dx\right)$. As a result of the imposition of the stress, the point which was originally at x displaces by an amount u whereas the point which was at $x + dx$ displaces by an amount $\left(u + \frac{\partial u}{\partial x} \cdot dx\right)$. The strain in the element ϵ therefore equals $\frac{u + \frac{\partial u}{\partial x} \cdot dx - u}{\partial x} = \frac{\partial u}{\partial x}$.

Assuming that the bar remains within its elastic region in which $\sigma = E\epsilon$, where E is Young's modulus for the material, the stress associated with the strain $\frac{\partial u}{\partial x}$ is

$$E \cdot \frac{\partial u}{\partial x} \tag{6.1}$$

The resultant force on the element is $\left(A \frac{\partial \sigma}{\partial x} \cdot dx\right)$ which, on substitution from Equation 6.1, becomes

$$AE \frac{\partial^2 u}{\partial x^2} dx$$

Writing down Newton's second law for the element gives

$$AE \frac{\partial^2 u}{\partial x^2} dx = \rho A dx \cdot \frac{\partial^2 u}{\partial t^2}$$

i.e. $$\frac{\partial^2 u}{\partial t^2} = \frac{E}{\rho} \cdot \frac{\partial^2 u}{\partial x^2} \tag{6.2}$$

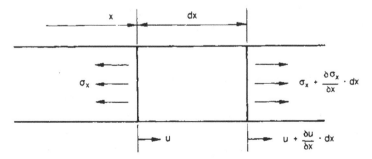

Figure 6.5 *Propagation of a stress wave in a solid bar*

The solution to this equation (which can be checked by substitution) is

$$u = a \sin b(x - \sqrt{\frac{E}{\rho}} \cdot t) \qquad (6.3)$$

where a and b are arbitrary constants. Equation 6.3 describes the way in which the displacement u of a point varies with time and position. For example, if we take a 'snapshot' view at a particular value of t, u varies sinusoidally with x and similarly u varies sinusoidally with time at a particular location x. In order to determine how fast the wave is travelling, we need to investigate the relationship between x and t for a particular value of u. Take, for example, the peak value of u, the peak of the wave, when $u = a$. For this value of u

$$b(x - \sqrt{\frac{E}{\rho}} \cdot t) = \frac{\pi}{2} + 2n\pi \quad (n = 1, 2 \ldots)$$

i.e. $\quad x = \frac{1}{b}(\frac{\pi}{2} + 2n\pi) + \sqrt{\frac{E}{\rho}} t$

Therefore

$$c = \frac{dx}{dt} = \sqrt{\frac{E}{\rho}} \qquad (6.4)$$

The propagation velocity of an elastic wave is the square root of the ratio of its elastic modulus to its density. By a similar argument it can be shown that a transient increment of stress, $\Delta\sigma$, in the plastic region (Figure 6.6) will propagate at a velocity $c = \sqrt{\frac{S}{\rho}}$ where S is the slope of the stress-strain curve at that level of stress. The conclusion we draw from this is that plastic waves travel less quickly than elastic waves.

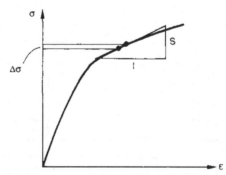

Figure 6.6 *Stress-strain curve for elasto-plastic material*

6.2.2 Particle velocity

A compressive stress σ is suddenly applied to the left hand end of a large bar of cross-sectional area A and density ρ. After a time t, the wavefront has travelled a distance $\sqrt{\frac{E}{\rho}} . t$ and it travels an extra distance $\sqrt{\frac{E}{\rho}} . \delta t$ during a time increment δt. During that time increment, the stress σ has delivered an impulse to the bar equal to $\sigma A \delta t$. This impulse has caused an increase in velocity, and therefore momentum, of each particle in the shaded element of Figure 6.7. i.e. $\sigma A \delta t = \rho A \sqrt{\frac{E}{\rho}} . \delta t . v$ where v is the particle velocity acquired. Whence

$$v = \frac{\sigma}{\sqrt{E\rho}} \tag{6.5}$$

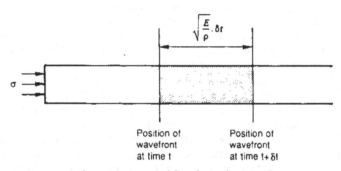

Figure 6.7 *Stress wavefront causing particle velocity increment*

Comparing Equations 6.4 and 6.5 reveals that, within the elastic region, the propagation velocity is solely determined by material properties whereas the particle velocity is proportional to the stress intensity of the wave. For plastic waves, the propagation velocity is indirectly affected by stress level since it is proportional to the square root of the slope of the stress-strain curve.

6.3 Reflection and transmission of stress waves

It has already been argued (Section 6.1) that a compressive wave will be reflected as a tensile wave (and vice versa) when it strikes a free surface. The wave system must be self-equilibriating since, by definition, no external force is available at a free surface. We next consider what happens when a stress wave is reflected from a fixed surface. The key feature of a fixed surface is that no displacement of the surface is possible. In order to visualise a fixed surface we consider a bar of relatively soft material (e.g. rubber) connected at one end to a rigid, steel abutment (Figure 6.8).

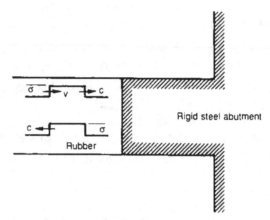

Figure 6.8 *Stress wave reflection at a fixed end*

A compressive pulse of intensity σ travels from left to right towards the fixed end, giving the molecules of the rubber bar a particle velocity from left to right (see Section 6.2). At the fixed surface the pulse will be reflected and travel back from right to left. During the time when the pulse is interacting with the fixed surface, no displacement and so no velocity of the rubber molecules adjacent to the abutment is permitted. The reflected wave must therefore have a particle velocity which will precisely negate the particle velocity of the incoming wave, i.e. of the same magnitude but opposite in direction. The reflected pulse, which is travelling from right to left, therefore has a particle velocity associated with it which is also directed from right to left. This we recognise as a compressive wave and we observe that compressive stress waves do not change sign when they reflect from a fixed surface. An exactly parallel argument can be developed for tensile waves, leading to the same conclusion. In summary, therefore, we can say that, when reflected from a free surface, a stress wave will change sign, but when reflected from a fixed surface, it will not. It is worth observing here that, for the duration of the interaction of the pulse with the fixed surface, the abutment experienced a stress of intensity 2σ and that a pulse of this intensity and duration will have been induced in the steel abutment.

6.3.1 Reflection and transmission at an interface between two dissimilar materials

The foregoing raises the question of what happens if the wave strikes neither a fixed nor a free surface but an interface between two dissimilar materials.

We assume a compressive wave of intensity σ_a strikes an interface between material 1, with properties ρ_1 and E_1, and material 2, with properties ρ_2 and E_2 (Figure 6.9). At this stage we do not know the sign of the transmitted and reflected waves σ_b and σ_c so we assume them to be compressive, adopting the convention that compressive stresses are positive and tensile stresses are negative.

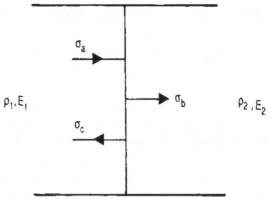

Figure 6.9 *Reflection and transmission at an interface between dissimilar materials*

The equilibrium condition at the interface yields

$$\sigma_a + \sigma_c = \sigma_b \qquad (6.6)$$

The compatibility condition states that the velocity of a particle to the left of the interface must equal the velocity of a particle to the right i.e.

$$\frac{\sigma_a}{\sqrt{E_1\rho_1}} - \frac{\sigma_c}{\sqrt{E_1\rho_1}} = \frac{\sigma_b}{\sqrt{E_2\rho_2}} \qquad (6.7)$$

Solving the simultaneous Equations 6.6 and 6.7 gives

$$\sigma_b = 2\left(\frac{\sqrt{E_2\rho_2}}{\sqrt{E_1\rho_1} + \sqrt{E_2\rho_2}}\right) \cdot \sigma_a \qquad (6.8)$$

and

$$\sigma_c = \left(\frac{\sqrt{E_2\rho_2} - \sqrt{E_1\rho_1}}{\sqrt{E_2\rho_2} + \sqrt{E_1\rho_1}}\right) \cdot \sigma_a \qquad (6.9)$$

Example 6.1
A composite armour system is constructed by laminating together a sheet of steel and a sheet of aluminium. A compressive wavefront of intensity 110 Nmm^{-2} travels through the steel plate and strikes the interface with the aluminium plate. Determine the intensity of the transmitted and reflected waves.

	$E(\text{Nm}^{-2})$	$\rho(\text{kg m}^{-3})$	$\sqrt{E\rho}$ (kg m^{-1}s^{-1})
Steel	20×10^{10}	8000	40×10^6
Aluminium	8×10^{10}	2800	15×10^6

Solution

Taking steel as material 1 and aluminium as material 2 and substituting in Equations 6.8 and 6.9 gives

$$\sigma_b = \frac{6}{11} \cdot \sigma_a = 60 \, \text{Nmm}^{-2}$$

and $\sigma_c = -\dfrac{5}{11}\sigma_a = -50 \, \text{Nmm}^{-2}$

A compressive wavefront of intensity 60 Nmm^{-2} is transmitted into the aluminium plate and a tensile wavefront of intensity 50 Nmm^{-2} is reflected back through the steel plate.

6.3.2   Reflection and transmission at a change of area

Another device by which a wave can be partially reflected is by changing the area of the medium through which it is transmitted. In practice this might be achieved by introducing cavities into the material, by employing a foamed or honeycomb construction, but we will consider the simplified representation of Figure 6.10.

The equilibrium condition at the interface yields

$$(\sigma_a + \sigma_c)A = \sigma_b \cdot nA \tag{6.10}$$

whilst the compatibility condition gives

$$\frac{\sigma_a}{\sqrt{\rho E}} - \frac{\sigma_c}{\sqrt{\rho E}} = \frac{\sigma_b}{\sqrt{\rho E}} \tag{6.11}$$

Solving Equations 6.10 and 6.11 gives

$$\sigma_b = \frac{2}{n+1} \cdot \sigma_a \tag{6.12}$$

and $\sigma_c = \dfrac{n-1}{n+1} \cdot \sigma_a \tag{6.13}$

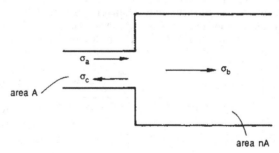

area A

σ_a

σ_c

σ_b

area nA

Figure 6.10   *Reflection and transmission at a change in area*

Example 6.2
A plastic, protective panel of lightweight, sandwich construction comprises two solid outer skins separated by a honeycombed core of the same material. The area of transmission in the core is estimated to be one-half that of the skins. Examine the reflections and transmissions which occur when a compressive stress of 18 Nmm^{-2} is suddenly applied to the face of the panel and maintained.

Solution
For the purpose of analysis, we can represent the panel by the system depicted in Figure 6.11.

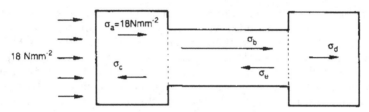

Figure 6.12 *Reflections and transmissions in a panel of sandwich construction*

At interface A, $\sigma_a = 18\,\text{Nmm}^{-2}$ and $n = \dfrac{1}{2}$

Equation 6.12 gives $\sigma_b = \dfrac{4}{3}\sigma_a = 24\,\text{Nmm}^{-2}$ (compressive) and

Equation 6.13 gives $\sigma_c = -\dfrac{1}{3}\sigma_a = -6\,\text{Nmm}^{-2}$ (tensile)

The wavefront transmitted at interface A is the source wavefront for interface B, where $n = 2$. Thus

$$\sigma_d = \frac{2}{3} \cdot 24 = 16\,\text{Nmm}^{-2}$$

and $\sigma_e = \dfrac{1}{3} \cdot 24 = 8\,\text{Nmm}^{-2}$

6.3.3 Reflection and transmission at an interface with change of area and change of material

For this case we adopt the equilibrium Equation 6.10 and the compatibility equation 6.7, yielding the following expression:

$$\sigma_b = \left(\frac{2\sqrt{E_2 \rho_2}}{\sqrt{E_1 \rho_1} + n\sqrt{E_2 \rho_2}} \right) \cdot \sigma_a \tag{6.14}$$

$$\sigma_c = \left(\frac{n\sqrt{E_2 \rho_2} - \sqrt{E_1 \rho_1}}{n\sqrt{E_2 \rho_2} + \sqrt{E_1 \rho_1}} \right) \cdot \sigma_a \tag{6.15}$$

6.3.4 Design of protective systems

The foregoing sections indicate the potential of a laminated composite for use as part of a protective system. By judicious selection of the components, a stress wave can be partially reflected at a series of interfaces, thus avoiding the large tensile stresses which can develop when a compressive stress reflects from the rear face of a monolithic target. It will be noted that the fraction of the wave which is reflected or transmitted at an interface is solely a function of material properties and transmission areas, therefore being independent of stress level. This means that a protective system does not need to be tuned to a specific level of attack.

If space and structural constraints permit, the inclusion of an air gap can prove a very effective method of eliminating scabbing off the rear face of a target. This method has been used in armoured fighting vehicles to offer protection against high explosive squash head rounds. A skirting plate, stood off from the main armour, detonates the round, thus dissipating much of its effect. Invariably scabs or fragments are produced from the skirting plate, but these have little effect on the main armour. Similar techniques have been adopted in structural applications where double walls and sacrificial roofs are used – the inner leaf catching the spall from the rear of the outer leaf.

If this is not possible, an estimate of the intensity of the stress will be necessary to determine whether or not a scab will be formed in the protective system. For inert projectiles, this can be effected by using either the Recht equation (see Section 12.6.1) or the first order hydrodynamic model (Section 12.6.2), depending on the nature of the attack. For in-contact charges, the applied pressure results from the detonation wave, discussed earlier in Section 2.5.3. A method is described in Chapter 7 for assessing the stresses transmitted to a buried structure from groundshock.

Example 6.3
Figure 6.12a shows a plane compressive pulse travelling through a plate of cast iron towards the free back surface of the plate. Calculate the thickness of the scab which will be produced on reflection at the back face of the plate. (For cast iron, $\rho = 7800$ kgm^{-3}, $E = 210$ kNmm^{-2}. A tensile stress of 250 Nmm^{-2} will cause fracture in the plate.)

Solution
The net tensile stress is given by the amount by which the reflected tensile wave exceeds the incoming compressive wave. Figure 6.12b shows the wave 2.5 μs after the leading edge reflects from the rear face of the target. The net tensile stress at this stage is 250 Nmm^{-2}, so a scab is formed. Since the propagation velocity of the wave (Equation 6.4) is 5189 m/s, it travels 13 mm in 2.5 μs. This is the thickness of the scab produced.

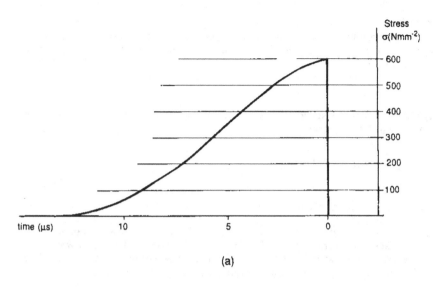

(a)

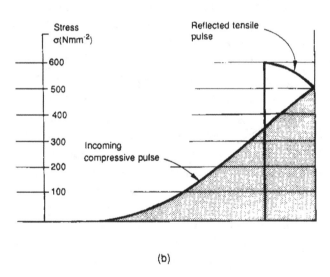

(b)

Figure 6.12 *Reflections and transmissions in a panel of sandwich construction*

6.4 X,T diagrams

Keeping track of the development of the stress wave pattern within a multi-layered target quickly becomes extremely difficult. The *x,t* diagram provides a systematic representation of the progress of a family of stress waves, enabling the user to identify the stress level at any point *x* in the system at a chosen time *t*.

Example 6.4

Draw an x,t diagram for the initial 100 μs period for the situation, described in Example 6.1. Take each plate to be 200 mm thick and assume the compressive pulse lasts for 20 μs.

Solution

Firstly we determine the speed of propagation of the waves in the steel (c_{st}) and the aluminium (c_{al}) (see Equation 6.4).

$$c_{st} = \sqrt{\frac{2 \times 10^{11}}{8 \times 10^3}} = 5000 \text{ ms}^{-1}$$

$$c_{al} = \sqrt{\frac{8 \times 10^{10}}{2.8 \times 10^3}} = 5345 \text{ ms}^{-1}$$

We now proceed to construct the x,t diagram by drawing a line to represent the progress of each wavefront (Figure 6.13). The line AB represents the initial wavefront σ_a, BC the transmitted wavefront σ_b, BD the reflected wavefront σ_c, and so on. The dotted line EF represents the tensile wavefront generated in terminating the pulse

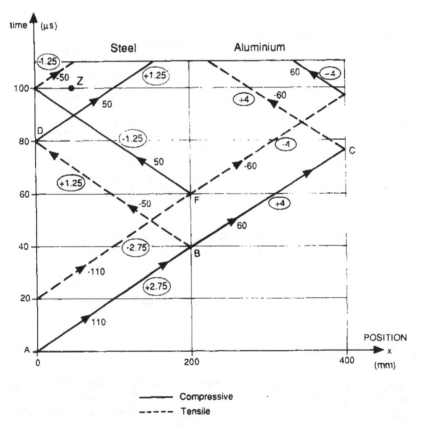

Figure 6.13   x, t *diagram for Example 6.4*

after 20 μs (see Section 6.2). The arrow heads on the lines represent the direction in which the wavefronts are propagating and the numbers next to the arrowheads indicate the stress level associated with each wavefront (compressive-positive). The numbers in the circles represent the particle velocities (in ms^{-1}) associated with each wavefront (see Equation 6.4) (here a positive value indicates the particles are given a velocity from left to right).

The stress level at a particular point and time can now be determined by summing up the effects associated with all the wavefronts which have passed that point since $t = 0$. Consider, for example, the point z, which represents the state at a point 50 mm from the face of the steel plate, 100 μs after the event commenced. The stress level is given by (working vertically upwards from $t = 0$)

$$\sigma_z = 110 - 110 - 50 + 50 + 50 = 50 \, \text{Nmm}^{-2} \quad \text{(compressive)}$$

and the particle velocity,

$$V_z = 2.75 - 2.75 + 1.25 - 1.25 + 1.25 = 1.25 \, \text{ms}^{-1} \quad \text{(from left to right)}$$

6.5 Plastic stress waves

It was argued in Section 6.2.1 that a plastic wave would travel less quickly than an elastic wave. We now examine the propagation and interaction of plastic stress waves in greater detail. Let us consider first a wave front of intensity 300 Nmm^{-2} initiated at time $t = 0$ at the left hand end of the bar shown in Figure 6.14a.

The stress-strain relationship for the material, whose density is 8000 kgm^{-3}, is the same in tension and compression and is described in Figure 6.14b. Using Equation 6.4, we discover that the velocity in the elastic region is 5000 ms^{-1}, but is only 2500 ms^{-1} in the plastic region. Figures 6.15a and b represent the changes in stress experienced at points A and B in Figure 6.14a.

These diagrams indicate that the further the observer is away from the source of the wave, the greater will be the delay between the arrival of the elastic and plastic components of the pulse. In other words the *shape* of the pulse is being modified as it travels down the bar. This becomes more evident if we now replace the bar of Figure 6.14a with one having the material characteristics shown in Figure 6.16.

Now each increment of stress above the yield point travels at a different speed. Figures 6.17a and b represent, schematically, the changes in stress observed at points A and B of Figure 6.14a.

Clearly the higher stress increments suffer increasing amounts of delay as the wave travels away from the source.

6.5.1 Reflection and transmission of plastic waves

The analysis of the interaction of plastic waves with boundaries and interfaces requires some intuition and a little inspired guesswork! Consider first a 1 metre long bar, with the material characteristics described in Figure 6.14b, fixed to a rigid abutment at its right hand end. A compressive stress of

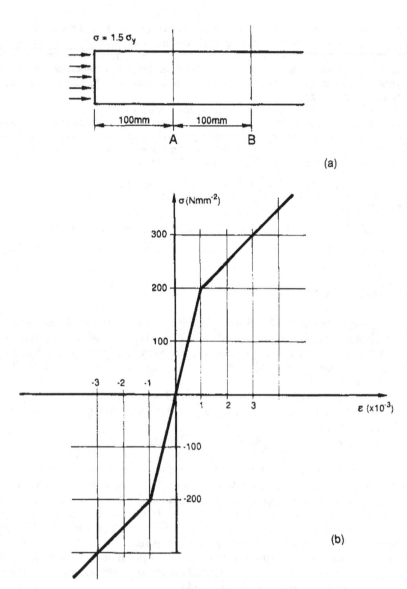

Figure 6.14   *Bilinear stress-strain relationship for elastic-plastic material*

intensity 150 Nmm^{-2} is applied to the left hand end of the bar and maintained. When the wavefront reaches the right hand end, which is fixed, it is reflected back as a compressive wavefront of the same intensity, 150 Nmm^{-2}. This overlies the compressive stress of 150 Nmm^{-2} laid down by the wavefront as it travelled down the bar, producing a total stress of 300 Nmm^{-2}. Since this exceeds the elastic limit, an elastic wavefront of 50 Nmm^{-2} and a plastic wavefront of 100 Nmm^{-2} are generated (see Figure 6.18).

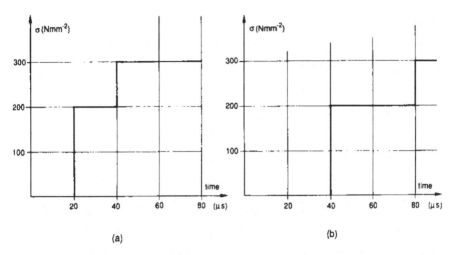

Figure 6.15 *(a) Stress variation with time at location A; (b) stress variation with time at location B*

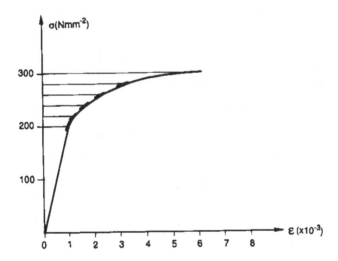

Figure 6.16 *Typical elasto-plastic material behaviour*

Example 6.5

A compound bar is constructed by joining end to end two bars, bar B having 60% of the area of bar A. A compressive stress wave of intensity 210 Nmm^{-2} travels from right to left through the composite bar shown in Figure 6.19. The bar is made of a material with the uniaxial stress strain characteristics shown in Figure 6.20. Determine the intensity of the transmitted and reflected wave.

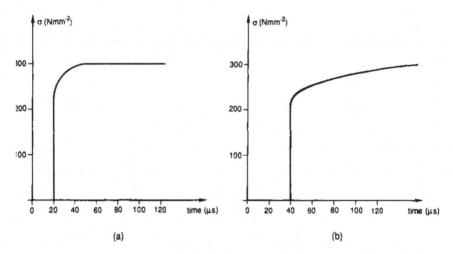

Figure 6.17 *(a) Stress variation with time at location A; (b) stress variation with time at location B*

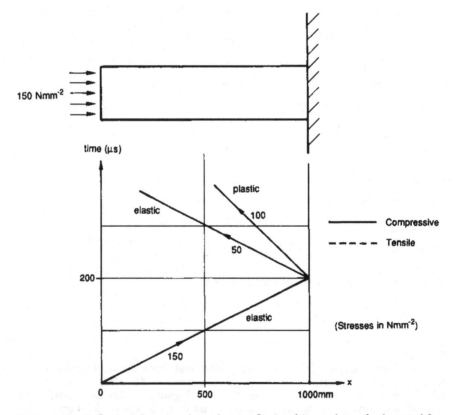

Figure 6.18 *Reflection of compressive pulse at a fixed end in an elasto-plastic material*

Figure 6.19 *Compound bar*

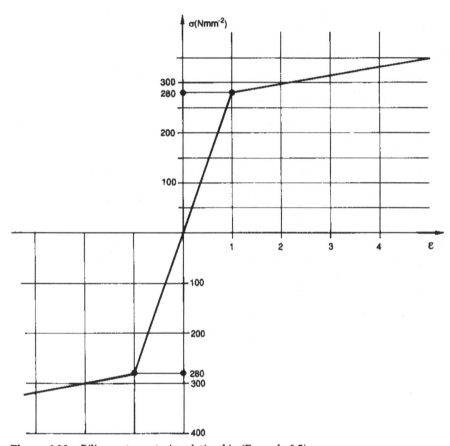

Figure 6.20 *Bilinear stress-strain relationship (Example 6.5)*

Solution
We start off by assuming that both elements of the bar remain elastic, in which case, using Equations 6.12 and 6.13,

$$\sigma_b = +336 \, \text{Nmm}^{-2} \, \text{(compressive)} \quad \text{and} \quad \sigma_c = -126 \, \text{Nmm}^{-2} \, \text{(tensile)}$$

In this case, the transmitted wave exceeds the yield strength of the material and so contradicts our assumption that the system remains elastic, implying our assumption was wrong.

From Figure 6.20 it can be seen that the gradient in the plastic region is $\frac{1}{16}$ of the gradient in the elastic region. The propagation velocity in the plastic region is therefore one-quarter of the elastic propagation velocity, c. We assume that, since the transmitted wave exceeds the yield stress of the material, both a plastic wave and an elastic wave, equal in intensity to the yield stress, are transmitted (Figure 6.21).

Considering equilibrium at the change of section gives

$$(210 + \sigma_c)4A = (280 + \sigma_b^P)A$$

i.e. $\sigma_b^P - 4\sigma_c = 560$ (6.16)

The compatibility condition gives

$$\frac{210}{\rho c} - \frac{\sigma_c}{\rho c} = \frac{280}{\rho c} + \frac{\sigma_b^P}{\rho 0.25c}$$

i.e. $4\sigma_b^P + \sigma_c + 70 = 0$ (6.17)

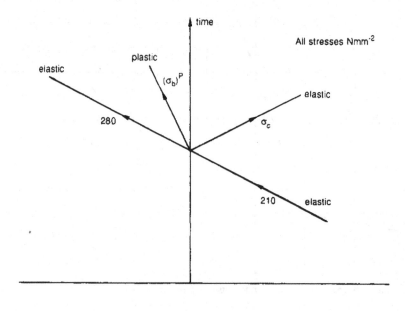

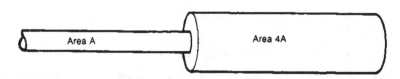

Figure 6.21   x, t *diagram for compound bar*

Solving equation 6.16 and 6.17 gives

$$\sigma_b^p = 16.5 \, \text{Nmm}^{-2} \quad \text{(compressive)}$$

$$\sigma_c = -135.9 \, \text{Nmm}^{-2} \quad \text{(tensile)}$$

In summary, a compressive elastic wave of intensity 280 Nmm^{-2} and a compressive plastic wave of 16.5 Nmm^{-2} are transmitted, while an elastic tensile wave of intensity 135.9 Nmm^{-2} is reflected.

Symbols

a	arbitrary constant
A	cross-sectional area of bar
b	arbitrary constant
c	wave propagation velocity
E	Young's modulus
E_1	Young's modulus of material 1
E_2	Young's modulus of material 2
F	force
n	factor by which bar area changes
S	slope of stress-strain curve in plastic region
t	time
u	displacement of point at position x
v	particle velocity
x	position of general point in a bar
ϵ	longitudinal strain
ρ_1	density of material 1
ρ_2	density of material 2
σ	uniaxial stress level
σ_a	intensity of incident wave
σ_b	intensity of transmitted wave
σ_b^p	intensity of transmitted plastic wave
σ_c	intensity of reflected wave

7 Groundshock

7.1 Introduction

Protection from aerially delivered weapons and airborne blast effects can be sought by going underground. However, buried structures can be vulnerable to transient stresses propagated through the soil and rock in which they have been constructed. Moreover, sensitive equipment, mounted in either surface-built or buried installations, may suffer damage from transmitted groundshock. This chapter provides the designer with the necessary information to predict the level of groundshock which will result from an event and to assess the potential damage which may be caused to a structure and its contents.

A description of wave propagation in geologic materials is presented together with empirical prediction methods for groundshock attenuation with range. Hence the groundshock loading of a buried structure can be predicted, from which the transient response of the structural elements can be determined. An approximate method of estimating gross structural movement will be presented and the methodology by which this can lead to quantification of potential damage to in-structure components will be discussed.

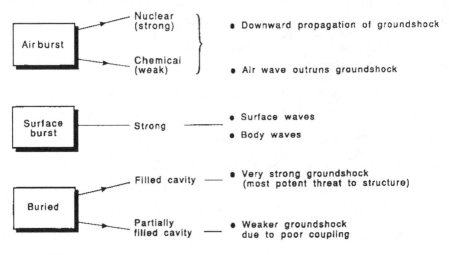

Figure 7.1 *Sources of groundshock*

7.2 Characterisation of groundshock

Figure 7.1 describes the principal sources of groundshock. Although a nuclear device, detonated above ground, can generate significant ground-shock levels, airburst of high explosive shells causes only weak levels of groundshock which attenuate quickly with range. The same weapon det-onating on the ground surface will generate much greater stresses due to the enhanced degree of coupling between the explosion and the ground. The most potent threat comes from the detonation of a weapon below ground, the effect being much greater when there are no air voids surrounding the weapon, resulting in a high level of coupling.

All explosive events near or under the ground's surface will cause both surface waves and body waves (Figure 7.2). The isotropic component of the transient stress pulse causes compression and dilation of the soil or rock with particle motions parallel to the direction of propagation of the wave (Figure 7.3). These are known as compression or 'P' waves. The deviatoric component of the stress pulse causes distortion, or shearing, of the soil with particle velocity perpendicular to the direction of propagation of the waves. These are known as shear or 'S' waves. Near the

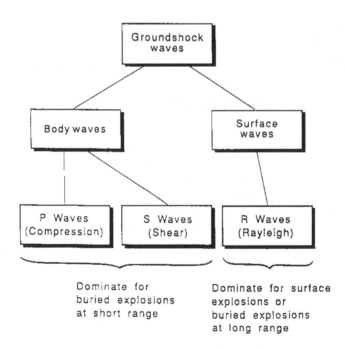

Figure 7.2 *Categorisation of groundshock*

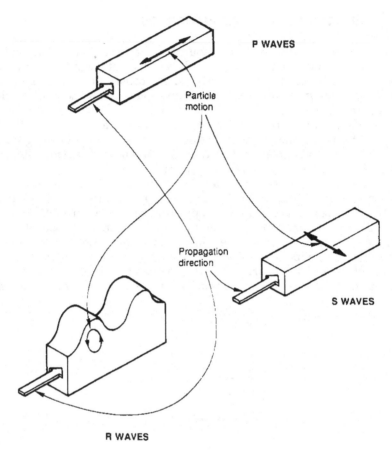

Figure 7.3 *Particle motions in P, R and S waves*

ground surface particles adopt a circular motion, similar to that experienced by a swimmer encountering waves in the sea. These are known as Rayleigh or 'R' waves. As a general rule P and S waves dominate for buried explosions at short range, whereas R waves dominate for surface explosions and, due to their slower rate of decay with distance, for buried explosions at larger ranges. The propagation velocities of both surface and body waves depend principally on the density and stiffness of the soil. R and S waves, which are both concerned with distortive movements in the soil, travel at approximately the same speed:

$$c_R \approx c_S = \sqrt{\frac{G}{\rho}} \tag{7.1}$$

where G is the shear modulus of the soil.

Since the propagation of P waves is concerned with the isotropic compression of soil

$$c_p = \sqrt{\frac{K}{\rho}}$$

where K is the bulk modulus and is given by

$$\frac{2}{3}G\frac{(1+\nu)}{(1-2\nu)} \tag{7.2}$$

The general term seismic velocity c is defined as follows

$$c = \sqrt{\frac{E}{\rho}} \tag{7.3}$$

where E is the modulus obtained from a uniaxial, unconfined compressive test. Seismic velocities vary from values of less than 200 m/s for loose, dry sand to values in excess of 1500 m/s for saturated clays. Wave energy and amplitude decrease with distance from the explosion for two reasons (Figure 7.4). Firstly, due to the geometric effect, the energy in the transient pulse is being spread over an increasing surface area as the spherical wavefront travels away from the site of the explosion. Secondly, energy is being dissipated in the soil as work is done in plastically deforming the soil matrix. A high seismic velocity normally implies low hysteresis and therefore little hysteretic attenuation with range. The amplitude of the body waves emanating from an event is approximately inversely proportional to the range from the event. The amplitude of surface waves, however, is inversely proportional to the square root of the range. P and S waves are attenuated more rapidly than R waves and so R waves tend to dominate at large range.

(1) GEOMETRIC EFFECT

(2) HYSTERESIS EFFECT

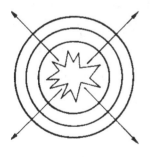

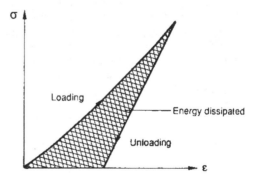

Figure 7.4 *Causes of attenuation*

7.3 Quantification of groundshock parameters

The magnitude of a transient disturbance is normally described by the following parameters:

x – peak particle displacement
u – peak particle velocity

It will be necessary to translate this statement of soil particle movement into the loading which the soil delivers to a buried structure or structural element. This is described by:

p_s – the peak side on overpressure
i_s – the side-on specific impulse

DETERMINING STRUCTURAL LOADING

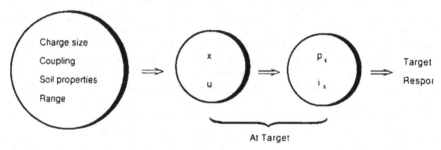

Figure 7.5 *Prediction of target response*

Figure 7.5 illustrates the procedure by which target response can be predicted.

The magnitude of the event is characterised by the yield of the explosion which is adjusted by a coupling factor to represent the extent to which the energy produced by the event is delivered into the soil in the form of groundshock. A knowledge of the soil properties and attenuation with range will allow an estimate to be made of x and u at the target. These are directly related to the loading parameters p_s and i_s, which can then be used to determine the target response in the same way as for airborne blast loading. The relationships between x, u, p_s and i_s are as follows:

$$p_s = \rho c_p u \tag{7.4}$$

and

$$i_s = \rho c_p x \tag{7.5}$$

7.4 SHM analogy

Some useful results are obtained by establishing an analogy between groundshock parameters and the simple harmonic motion (SHM) model of behaviour (Figure 7.6).

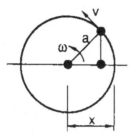

Figure 7.6 *The SHM analogy*

The projection of a point describing a circle of radius a at angular velocity ω onto the diameter of the circle executes SHM with amplitude a and frequency f, where $\omega = 2\pi f$. The amplitude of the oscillation a corresponds to the peak particle displacement x. The peak velocity of the oscillation, corresponding to the peak particle velocity u, is given by

$$v = a\omega = 2\pi f a \tag{7.6}$$

thus by analogy

$$u = 2\pi f x$$

$$\text{or } f = \frac{u}{2\pi x} \tag{7.7}$$

Equation 7.7 provides a value for the frequency of the groundshock disturbance and since $c = f\lambda$, the wavelength λ is given by

$$\lambda = \frac{2\pi cx}{u} \tag{7.8}$$

The use of the SHM analogy has enabled us to assign values of frequency and wavelength to a shock using only the values of u, x and c.

Example 7.1
Measurements taken of surface wave propagation, near a proposed site for a buried structure, indicate a seismic velocity of 300 m/s. The peak particle velocity, measured in metres per second, is found to be 250 times as large as the peak particle displacement measured in metres. Deduce the apparent frequency and wavelength of the disturbance and use Figure 7.7 to predict the depth at which the vertical component of the amplitude falls to 20% of its surface value.

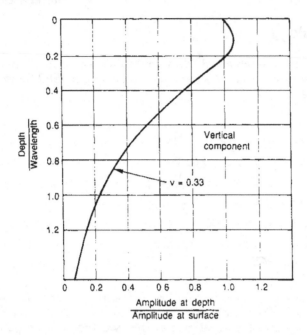

Figure 7.7   *Variation of amplitude with depth-surface waves*

Solution
From Equation 7.7

$$f = \frac{1}{2\pi} \cdot \frac{u}{x}$$

$$= \frac{250}{2\pi} = 39.8 \text{ Hz}$$

and $\lambda = \dfrac{c}{f} = \dfrac{300}{39.8} = 7.54$ m

From Figure 7.7, vertical component falls to 20% at

$$\frac{z}{\lambda} = 1$$

i.e. when $z = 7.54$ m. This example demonstrates the fact that surface waves are, indeed, confined to a relatively thin surface layer.

7.5   Groundshock predictions

The issue of groundshock attenuation with range was discussed in general terms in Section 7.2. All of the major parameters – peak pressure, peak particle velocity, peak particle displacement, specific impulse – reduce with distance from the event. The rate of attenuation with range is governed by soil type and characterised by an attenuation factor n, defined in Section 7.5.5. A large number of trials, conducted principally in the USA and detailed in Reference 1, have resulted in empirical relationships which permit predictions of the groundshock parameters. The following predictions relate to 'near field' events with a scaled distance of 5 m/kg$^{1/3}$ or less, in which body waves dominate.

7.5.1 Peak particle displacement

The peak particle displacement x caused by a totally or partially buried bomb at a location a distance R from the bomb can be estimated using Equation 7.9

$$\frac{x}{W^{1/3}} = 60 \cdot \frac{f_c}{c} \left(\frac{2.52R}{W^{1/3}}\right)^{1-n} \tag{7.9}$$

where x is measured in metres, W is the charge mass in kg, f_c is a dimensionless coupling factor (see Section 7.5.6), c is the seismic velocity in m/s, R is the range in metres, and n is a dimensionless attenuation coefficient (see Section 7.5.5).

7.5.2 Peak particle velocity

The peak particle velocity u at a range R from the bomb is given by Equation 7.10

$$u = 48.8 f_c \left(\frac{2.52R}{W^{1/3}}\right)^{-n} \tag{7.10}$$

7.5.3 Free field stress

The peak pressure p_0 generated in a free field environment is given by

$$p_0 = \rho \cdot C \cdot u \tag{7.11}$$

where ρ is the density of the soil kg/m^3, C is the loading wave velocity (defined below), u is peak particle velocity (determined from Equation 7.10), and p_0 is measured in N/m^2.

The loading wave velocity C depends on both the seismic velocity and the peak particle velocity. At short range, C is high due to the high values of particle velocity, but its value decays to the seismic velocity as range increases.

The value adopted for C should never be less than c and is given by

$$C = c \quad \text{for fully saturated clays}$$

$$C = 0.6c + \left(\frac{n+1}{n-2}\right)u \quad \text{for saturated clays}$$

$$C = c + \left(\frac{n+1}{n-2}\right)u \quad \text{for sand}$$

where n is the attenuation coefficient (see Section 7.5.5).

7.5.4 Free field impulse

The specific impulse is evaluated from Equation 7.12

$$i_o = \rho C x \tag{7.12}$$

where ρ is the density (kg/m^3), x is the peak particle displacement (Equation 7.9), C is the loading wave velocity (see Section 7.5.3), and i_o is measured in Ns/m^2.

7.5.5 Attenuation coefficient

The prediction methods presented in Sections 7.51 to 7.54 are based on field data of the type illustrated in Figure 7.8. The gradient of the line indicates the rate at which attenuation is taking place. In this case the

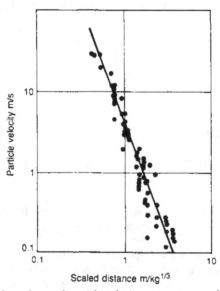

Figure 7.8 *Typical dependence of particle velocity upon range and charge weight*

Table 7.1

Soil type	Attenuation coefficient (n)
Saturated clay	1.5
Partially saturated clay and silt	2.5
Very dense sand, dry or wet	2.5
Dense sand, dry or wet	2.75
Loose sand, dry or wet	3.0
Very loose sand, dry or wet	3.25

gradient is approximately -2.8 and the attenuation coefficient n is 2.8. Some typical values of the attenuation coefficient are given for a range of soils in Table 7.1.

For soils not mentioned in Table 7.1 an estimate of the attenuation coefficient can be made from an unconfined uniaxial compression test, the results of which are represented schematically in Figure 7.9. If ϵ_0 represents the residual strain after loading to a stress level representative of the ground shock,

$$n = \frac{2 + \epsilon_0}{1 - \epsilon_0} \tag{7.13}$$

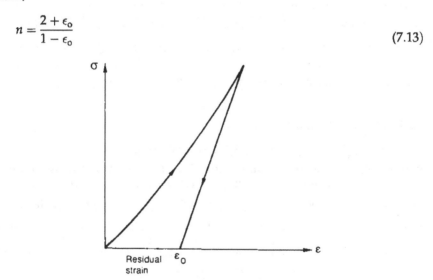

Figure 7.9 *Typical result obtained in a uniaxial compression test*

7.5.6 Coupling factor

The effectiveness of a weapon increases with the degree of coupling between the explosion and the ground, which is a function of its depth of burial. This effect is quantified by the incorporation of the coupling factor, f_c , in equations 7.9 and 7.10. Values for the coupling factor are given in Figure 7.10.

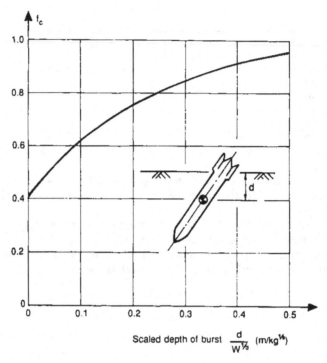

Figure 7.10   *Dependence of coupling factor on scaled depth of burst*

7.5.7   Westine's method

An alternative groundshock prediction method, due to Westine,[2] covers a wider range of scaled distance. Stage 1 is to modify the actual energy released by the event E_a to account for the degree of coupling between the explosive and the ground. The parameter $\frac{E_a}{P_a r^3}$ is evaluated for the event in which E_a is the energy released in Joules, P_a is the atmospheric pressure in N/m^2 and r is the radius of a sphere having the same volume as the underground cavity in which the explosive is located. Entering Figure 7.11 gives a value of the parameter $\left(\frac{E_e}{E_a}\right)\left(\frac{\rho c_p^2}{P_a}\right)^{0.760}$, in which ρ is the density of the soil (kg/m^3) and c_p is the P wave velocity. Hence a value of E_e, the equivalent fully coupled energy release, is determined.

Stage 2 is to use Figures 7.12 and 7.13 to obtain values for peak particle displacement x and peak particle velocity u. Stage 3 is to translate these into pressure and specific impulse values using Equations 7.4 and 7.5.

Example 7.2
A 7 kg mass of TNT is detonated in a cavity of diameter 6 m in a soil of density 2000 kg/m^3 having c_p = 300 m/s. Use the Westine method to predict the groundshock at a distance of 20 m from the centre of the explosion.

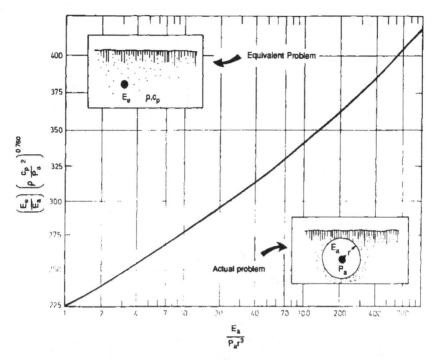

Figure 7.11 *Equivalent effective energy release (after Ref. 2)*

Solution
The energy release per kg of TNT = 4520 kJ/kg, so $E_a = 7 \times 4520 = 31640$ kJ.

Thus $\qquad \dfrac{E_a}{P_a r^3} = \dfrac{3.164 \times 10^7}{10^5 \cdot 27} = 11.72$

Entering Figure 7.11 gives $\qquad \left(\dfrac{E_e}{E_a}\right)\left(\dfrac{\rho c_P^2}{P_a}\right)^{0.76} = 282$

i.e. $E_e = 0.95\, E_a = 30\,058$ kJ, so the parameter

$$\frac{E_e}{\rho c_P^2 R^3}$$

takes the value 2.1×10^{-5}.

Entering Figure 7.12 gives $\qquad \left(\dfrac{x}{R}\right)\left(\dfrac{P_a}{\rho c_P^2}\right)^{1/2} = 3.05 \times 10^{-7}$

which gives $x = 2.6 \times 10^{-4}$ m and, from Figure 7.13

$$\left(\frac{u}{c_P}\right)\left(\frac{P_a}{\rho c_P^2}\right)^{1/2} = 8.3 \times 10^{-7}$$

which gives $u = 1.06 \times 10^{-2}$ m/s.

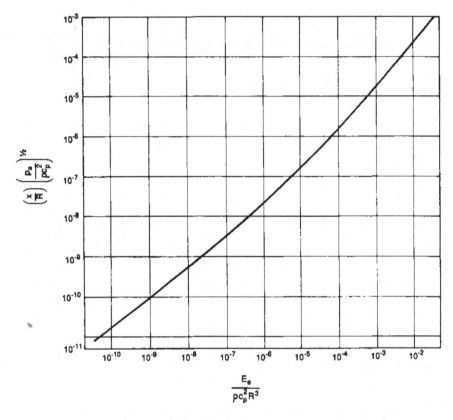

Figure 7.12   *Peak particle displacement values (after Ref. 2)*

7.6   Groundshock loading of structures

In order that a prediction of a structure's response to groundshock can be made, assumptions have to be made about the transient pressure pulse on the structure. As with blast waves, the pressure experienced by a structure during reflection is greater than the free-field pressure. Reference 3 recommends that free-field pressure values are multiplied by a factor of 1.5 to give reflected overpressures. The time for which the reflected overpressure acts on a particular point P on the structure (Figure 7.14) is determined by the time taken for a tension wave to propagate from a free edge to the point on the structure, thereby relieving the compressive reflected overpressure.

This is given by (Figure 7.14)

$$t_r = \frac{1}{c}\left(l_2 + \frac{l_3}{2} - l_1\right) \tag{7.14}$$

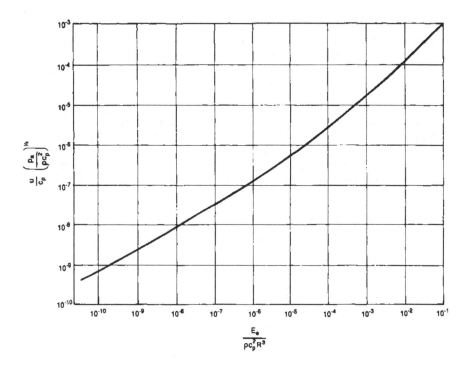

Figure 7.13 *Peak particle velocity values (after Ref. 2)*

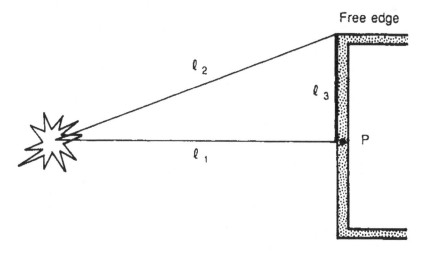

Figure 7.14 *Propagation path lengths for overpressure relief*

The transient pressure experienced at point P is represented by the simplified pulse shown in Figure 7.15, in which

$$t_d = \frac{2i_o}{p_o} \tag{7.15}$$

where i_o and p_o are free-field values of specific impulse and pressure

In practice the reflected overpressure will clear almost instantaneously at a location near a free edge, whereas t_r will be significant for a point near the centre of a wall. By convention, the transient pulse is simplified still further to that shown in Figure 7.16. Values of t_r and t_d are determined for a rectangular wall or roof slab at the point indicated in the diagram. This

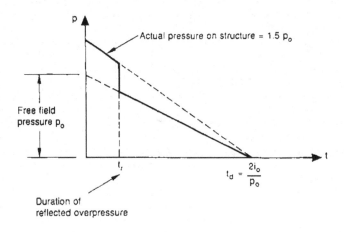

Figure 7.15   *Transient pressure experienced at point P (Figure 7.14)*

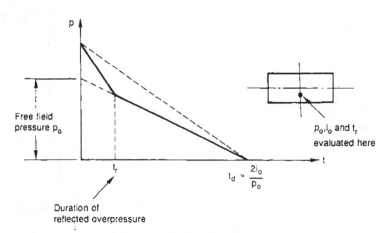

Figure 7.16   *Simplified representation of transient pulse*

pulse is used as a description of the transient load on a buried structural element and is utilised in the SDOF analysis of the element to assess structural response to groundshock (see Chapter 8).

7.7 In-structure shock

When groundshock strikes a buried installation vibrations will be generated within the fabric of the structure. Equipment attached to the walls or floor of the building will oscillate under the influence of these imposed motions. This section describes a simplified method for predicting these motions and presents a method by which potential damage to components within a structure can be assessed.

7.7.1 Wall and floor motions

Reference 1 gives the following expressions to describe the horizontal motions of a vertical wall (Figure 7.17)

$$x_w \leq 2x$$

$$u_w = 2u$$

$$a_w = \frac{2p_0}{m} \tag{7.16}$$

where x_w, u_w, a_w are the displacement, velocity and acceleration of the wall and x, u and p_0 are the free field values of displacement, velocity and pressure associated with the groundshock.

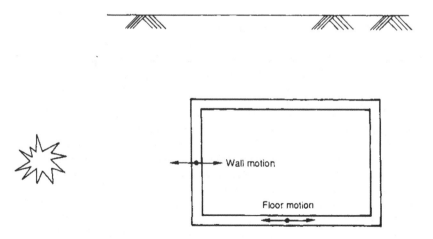

Figure 7.17 *Horizontal motions of wall and floor*

Horizontal floor motions due to side burst may be taken to equal the free-field values as follows

$$x_f = x$$

$$u_f = u \tag{7.17}$$

where x_f and u_f are the horizontal displacement and velocity of the floor.

7.7.2　*Equipment motions*

An item of equipment, together with the supports by which it is mounted on the wall or floor of a structure, constitutes a system, the oscillations of which are governed by:

1　The system's inertia.
2　The stiffness of the mounting.
3　The damping provided by the mounting.
4　The excitation provided by wall or floor motions.

Such a system can normally be analysed as a single degree of freedom system, from which a prediction of the peak displacement, velocity and acceleration of the oscillations can be made. For equipment with mountings providing damping in the range of 5–10% of critical, approximate values are given by:

$$x_s = 1.2x_w$$

$$u_s = 1.5u_w$$

$$a_s = 2a_w \tag{7.18}$$

where x_s, u_s and a_s are peak displacement, velocity and acceleration of the equipment.

7.7.3　*Assessing potential damage*

Having determined the severity of the vibration to which a piece of equipment will be subjected, it is now necessary to compare this with the maximum it can tolerate before incurring damage. A programme of trials must be established in which the item of equipment under scrutiny is subjected to a range of vibrations to determine the amplitude necessary to cause damage at a particular frequency of oscillation. This damage envelope can be plotted on 'tripartite' paper which simply takes advantage of the SHM relationships to provide a comprehensive presentation. Any two of the four variables – peak displacement, velocity, acceleration and frequency – determine the other two and completely defines the oscillating motion. Figure 7.18 is an example of the form which a damage envelope takes. The predicted motion from a groundshock event gives rise to a single point (P) on the diagram. If this point falls below the damage envelope the system will survive. If, however, the point falls above the envelope it will sustain damage. A

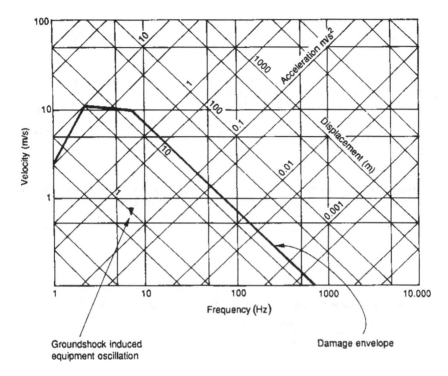

Figure 7.18 *Typical damage envelope*

more comprehensive description of in-structure shock analysis is given in the book by Dowding (see Bibliography).

7.8 Transmission and Reflection

When a groundshock pulse meets an interface between two dissimilar soils, both transmission and reflection from the interface will occur. The problem is similar to that of a stress wave striking an interface between two transmission media, a topic which was discussed in Section 6.3.1. The intensity of the transmitted shock, σ_b, and the reflected shock σ_c can be expressed in terms of the incident shock, σ_a, as follows

$$\sigma_b = 2\left(\frac{\rho_2 c_2}{\rho_1 c_1 + \rho_2 c_2}\right) \cdot \sigma_a$$

$$\sigma_c = \left(\frac{\rho_2 c_2 - \rho_1 c_1}{\rho_2 c_2 + \rho_1 c_1}\right) \cdot \sigma_a \tag{7.19}$$

where ρ_1 and ρ_2 are the densities of soils one and two and c_1 and c_2 are the propagation velocities in the two soils.

Significant benefits can be derived from surrounding a buried structure with loose sand or gravel as will be demonstrated in Example 7.3.

Example 7.3

A shock wave of intensity 6×10^4 N/m^2 is travelling horizontally through a saturated clay mass, which has a density of 2000 kg/m^3 and a seismic velocity of 2100 m/s. It strikes a vertical infill of poorly graded, loose sand of density 1600 kg/m^3 and a seismic velocity of 180 m/s. The infill surrounds a concrete structure, for which $\rho = 2300$ kg/m^3 and the seismic velocity is 3020 m/s. Assess the degree of groundshock insulation provided by the granular backfill.

Solution

If the sand infill were not present, the wave would pass directly from clay to concrete. The intensity of the wave transmitted into the concrete structure would be

$$\sigma_b = 2 \left(\frac{2300 \cdot 3020}{2000 \cdot 2100 + 2300 \cdot 3020} \right) \cdot 6 \times 10^4$$

$$= 7.5 \times 10^4 \text{ N/m}^2$$

However, when the wave passes from clay to sand the intensity is reduced

$$\sigma_b = 2 \cdot \left(\frac{1600 \cdot 180}{2000 \cdot 2100 + 1600 \cdot 180} \right) \cdot 6 \times 10^4$$

$$= 7.7 \times 10^3 \text{ N/m}^2$$

The wave transmitted into the concrete structure is now

$$\sigma_b = 2 \left(\frac{2300 \cdot 3020}{1600 \cdot 180 + 2300 \cdot 3020} \right) \cdot 7.7 \times 10^3$$

$$= 1.48 \times 10^4 \text{ N/m}^2$$

The inclusion of the granular backfill has reduced the stress transmitted into the concrete by 80%.

7.9   Concluding remarks

This chapter has presented methods for predicting the intensity of ground-shock disturbances at a known range from a buried or partially buried explosion. The loading which a buried structure will experience and the assessment of damage to equipment within a buried structure have been discussed. The less the degree of coupling between bomb and ground and the further the structure is from the explosion, the smaller will be the groundshock effects. Thus, in order to ensure maximum protection for a buried structure, a protective overlay of rock or concrete at the ground surface should be emplaced to resist the penetration of the bomb and the structure should be buried as deeply as possible. Where a structure is being constructed in a dense soil with high seismic velocity (e.g. clay) a buffer layer of loose granular soil (e.g. sand) around the structure will reduce the transmitted shock levels.

When a bomb is detonated in or near the surface of the ground a crater will be formed, from which ejecta may travel considerable distances, causing damage to buildings, equipment and personnel. Empirical information on the diameter and depth of the crater formed and the size and throw of debris is given in Reference 3.

7.10 References

[1] *Protective Construction Design Manual*. Report No. ESL-TR-87-57, US Air Force Engineering and Services Center, Tyndall Air Force Base, Florida
[2] Westine P.S., Groundshock from the Detonation of Buried Explosives. *Journal of Terramechanics*, Vol. 15, No. 2, (1978)
[3] Fundamentals of Protective Design for Conventional Weapons. *US Army Technical Manual No. 5-855-1*. Headquarters, Department of the Army, Washington DC, 1986

7.11 Bibliography

Dowding C.H., *Blast Vibration Monitoring and Control*. Prentice-Hall, Hemel Hempstead (1985)
Ajaya Kumar Gupta, *Response Spectrum Method*. Blackwell Scientific Publications, Oxford (1990)

Symbols

a_w	horizontal acceleration of vertical wall
a_s	peak acceleration of equipment
c	seismic velocity
C	loading wave velocity
c_P	propagation velocity of P waves
c_R	propagation velocity of R waves
c_S	propagation velocity of S waves
c_1	seismic velocity in soil 1
c_2	seismic velocity in soil 2
E	Young's modulus
E_a	actual energy released from a bomb
E_e	effective, fully coupled, energy release
f_c	coupling factor
f	frequency
G	shear modulus
i_o	free-field value of specific impulse
i_s	side-on specific impulse
K	bulk modulus

$l_1 \, l_2 \, l_3$	dimensions (Figure 7.14)
n	attenuation factor or coefficient
m	mass per unit area of structure
p_o	free-field value of groundshock overpressure
P_a	atmospheric pressure
P_s	peak side-on overpressure
x	peak particle displacement
x_w	peak displacement of wall
x_f	peak displacement of floor
x_s	peak displacement of equipment
r	radius of spherical cavity
R	range from event
t_d	duration of pulse
t_r	duration of reflected pressure
u	peak particle velocity
u_w	peak velocity of wall
u_f	peak velocity of floor
u_s	peak velocity of equipment
v	velocity of point describing circle in SHM
W	charge mass
z	depth below ground surface
ϵ_o	residual strain
ν	Poisson's ratio
ρ	density
ρ_1	density of soil 1
ρ_2	density of soil 2
σ_a	peak incident stress
σ_b	peak transmitted stress
σ_c	peak reflected stress
σ_s	peak stress in sand backfill
ω	angular velocity

8 Structural response: principles

8.1 Introduction

The response of structures to dynamic loads has been studied extensively and a large number of texts are available which deal with the subject in a comprehensive manner. Examples include Thompson,[1] Irvine,[2] Berg[3] and Paz.[4] One of the earlier works containing a complete treatment of the subject by Biggs[5] has unfortunately been unavailable for a number of years. The approach presented in this latter work has formed the basis for the principal code for the design of structures to resist blast and other dynamic loads, the document TM5-1300 'Structures to Resist the Effects of Accidental Explosions'.[6]

The approach presented in TM5-1300 is centred on the response of real distributed mass structures and structural elements which are idealised as 'equivalent lumped-mass single degree of freedom' systems. For this reason, this chapter will concentrate solely on the response of such systems, developing the response of an idealised single degree of freedom (SDOF) structure to an idealised blast load.

8.2 Structural vibrations

A structure can be disturbed from its normal position of static equilibrium in a number of ways. For instance, the structure can be given an initial displacement (i.e. moved from its static equilibrium position and released) after which it will vibrate with a period of oscillation dependent on its mass and a factor characterising the resistance that it develops. For an elastic structure, resistance can be characterised by the stiffness of a spring. The amplitude of vibration at a particular time after motion begins will be determined by the amount of damping to which the structure is subject. Damping forces are often found to be proportional to the instantaneous velocity of the structure. The alternative way of disturbing a structure is to give it an initial velocity (derived from an impulse applied to the structure) after which it will again vibrate as described above. These two responses are referred to as *free* vibration.

When a structure is caused to vibrate by the application of time-dependent loading, the response is termed *forced* vibration. In the context of this book the so-called forcing function will be a representation of the blast load exciting the structure. A special case of loading arises if the frequency of the applied loading matches the natural response frequency of the structure. In this case the resultant amplitude of vibration theoretically will be infinite. This effect is known as *resonance*.

8.3   Single degree of freedom systems

A general single degree of freedom damped system can be represented as shown in Figure 8.1.

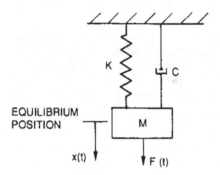

Figure 8.1   *Single degree of freedom elastic system with damping*

Here M is the structure mass, K is structure resistance: this may be an elastic resistance, as indicated here, or a more complex form of resistance. The term c is the structure damping coefficient: this factor will attenuate amplitude of vibration with time. There are a number of types of damping function. For example, in Coulomb or dry friction damping the damping force is given by $\pm \mu N$, where μ is the coefficient of friction between the structure and the surface on which it moves and N is the normal reaction force which may equal the weight of the structure. In hysteresis, damping force is proportional to the displacement of deformed material. This is also known as solid or structural damping. Finally, and most commonly, in viscous damping, force is proportional to the structure's velocity and is given by $c\dot{x}$. This is the most usual form of damping associated with blast loaded structures and will be discussed below in more detail.

Referring again to Figure 8.1, $F(t)$ is the time varying applied loading and may be written as $F[f(t)]$, where F is peak applied force, $f(t)$ is the time variation of the load and $x(t)$ is the time-dependent displacement of the structure. A free body diagram of the structure showing the forces acting is shown in Figure 8.2.

The equation of motion of this structure is

$$M\ddot{x} + c\dot{x} + Kx = F(t) \tag{8.1}$$

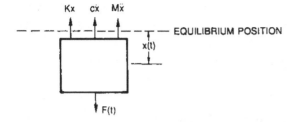

Figure 8.2 *Free body diagram of single degree of freedom system*

8.4 Free vibration

In the case of free vibration, which is produced when the structure is displaced from its equilibrium position and released or given an initial velocity, the forcing function $F(t) = 0$. Thus the equation of motion reduces to

$$M\ddot{x} + c\dot{x} + Kx = 0 \qquad (8.2)$$

In the case of an undamped vibration $(c = 0)$ the equation of motion further simplifies to

$$M\ddot{x} + Kx = 0 \qquad (8.3)$$

This can be written as

$$\ddot{x} + \omega^2 x = 0 \qquad (8.4)$$

where $\omega = \sqrt{\frac{K}{M}}$, the natural circular frequency of vibration in radians per second. The solution to this equation may be written

$$x(t) = A \sin \omega t + B \cos \omega t \qquad (8.5)$$

When $t = 0$ the displacement of the structure is $x(0) = x_0$ and so $x_0 = B$. Differentiating Equation 8.5 with respect to time gives

$$\dot{x}(t) = A\omega \cos \omega t - B\omega \sin \omega t \qquad (8.6)$$

and when $t = 0$ velocity $\dot{x}(0) = \dot{x}_0$ so $\dot{x}_0 = A\omega$ and therefore $A = \frac{\dot{x}_0}{\omega}$. The complete solution to the equation of motion is

$$x(t) = \frac{\dot{x}_0}{\omega} \sin \omega t + x_0 \cos \omega t \qquad (8.7)$$

8.5 Damped vibration: c finite

The equation of motion for this situation is

$$M\ddot{x} + c\dot{x} + Kx = 0 \qquad (8.8)$$

This can be written as

$$\ddot{x} + \frac{c}{M}\dot{x} + \frac{K}{M}x = 0 \tag{8.9}$$

If a damping factor ζ is defined as

$$\zeta = \frac{c}{2M\omega} \tag{8.10}$$

where ω is as defined above, Equation 8.9 becomes

$$\ddot{x} + 2\zeta\omega\dot{x} + \omega^2 x = 0 \tag{8.11}$$

If a solution $x = e^{Dt}$ is assumed (where D is a constant) and this is substituted into Equation 8.11, the so-called 'characteristic equation' is formed which has two roots as shown below.

$$D^2 + 2\zeta\omega D + \omega^2 = 0$$

$$D = \frac{-2\zeta\omega \pm \sqrt{4\zeta^2\omega^2 - 4\omega^2}}{2}$$

$$= \omega\left[-\zeta \pm \sqrt{\zeta^2 - 1}\right] = -\zeta\omega \pm \omega\sqrt{\zeta^2 - 1} \tag{8.12}$$

The cases of interest here have ζ less than unity which leads to the 'complementary function' which, in the absence of a forcing function, is the complete solution for this problem, obtained as

$$x(t) = e^{-\zeta\omega t}\left[C_1 \sin \omega\sqrt{1 - \zeta^2}t + C_2 \cos \omega\sqrt{1 - \zeta^2}t\right] \tag{8.13}$$

When $\zeta = 0$ (i.e. $c = 0$) the system is undamped and the solution reduces to the result above for undamped free vibration. Details of this analysis are provided in Reference 1.

As noted above, this particular solution is valid only for $\zeta < 1$. The case where $\zeta > 1$ (overdamping) is of limited practical importance in the context of blast loaded structures. As Figure 8.3 below shows, such a situation is not a truly vibrational response. Instead the structure creeps slowly back to its equilibrium position with no oscillatory component. The bigger the value of ζ the longer will be the recovery period. Generally *underdamped* systems (where ζ is between about 0.1 and 0.4) are more significant in the assessment of the response of structures subjected to blast loads.

It is convenient to introduce the frequency of the damped system ω_d which is related to the natural undamped frequency ω thus:

$$\omega_d = \omega\sqrt{1 - \zeta^2} \tag{8.14}$$

Therefore Equation 8.13 may be written

$$x(t) = e^{-\zeta\omega t}[C_1 \sin \omega_d t + C_2 \cos \omega_d t] \tag{8.15}$$

To determine the coefficients C_1 and C_2, note that when $t = 0$

$$x(0) = x_o = C_2 \tag{8.16}$$

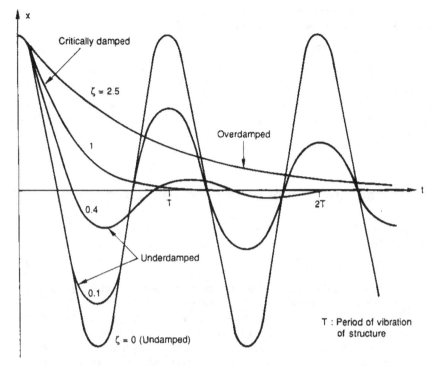

Figure 8.3 *Displacement vs time for SDOF systems with damping (after Ref. 3)*

Differentiation with respect to t gives

$$\dot{x}(t) = e^{-\zeta\omega t}[\omega_d C_1 \cos \omega_d t - \omega_d C_2 \sin \omega_d t] - \zeta\omega d^{-\zeta\omega t}[C_1 \sin \omega_d t + C_2 \cos \omega_d t]$$

$$= e^{-\zeta\omega t}[-(\zeta\omega C_1 + \omega_d C_2) \sin \omega_d t + (\omega_d C_1 - \zeta\omega C_2) \cos \omega_d t] \qquad (8.17)$$

and so when $t = 0$

$$\dot{x}(0) = \dot{x}_o = \omega_d C_1 - \zeta\omega C_2 \qquad (8.18)$$

Thus C_1 is evaluated as

$$\frac{C_1 = \dot{x}_o + \zeta\omega x_o}{\omega_d} \qquad (8.19)$$

and the complete solution is

$$x(t) = e^{-\zeta\omega t}\left[\frac{\dot{x}_o + \zeta\omega x_o}{\omega_d} \sin \omega_d t + x_o \cos \omega_d t\right] \qquad (8.20)$$

The special case where ζ is unity is known as critical damping and the damping coefficient is c_{crit}. Thus from Equation 8.10

$$c_{crit} = 2M\omega \qquad (8.21)$$

As ζ tends to unity so ω_d tends to zero and hence $\cos \omega_d t \approx 1$ and $\sin \omega_d t \approx \omega_d t$ which means that Equation 8.20 becomes

$$x(t) = e^{-\omega t}[\dot{x}_o t + (1 + \omega t)x_o] \tag{8.22}$$

The variation of displacement with time for a structure given an initial displacement of x_0 or an initial velocity $\dot{x}_0$ is shown in Figures 8.4a and b respectively for a typical underdamped system while the response of a critically damped system having an initial displacement is shown in Figure 8.4c.

8.6 Forced vibration of undamped structures (c = 0)

The equation of motion is now written to include a forcing function of the general form $Ff(t)$

$$M\ddot{x} + Kx = F(t) = Ff(t) \tag{8.23}$$

In order to assess the response of the structure now, consider a generalised forcing function of arbitrary history as shown in Figure 8.5.

Consider the area beneath the curve associated with the element of time $d\tau$. The area of this small block-pulse represents an impulse and will cause an increment in velocity of the system in accordance with Newton's second law which states that an impulse I produces a change in momentum (the product of mass × velocity change). Thus

$$I = \text{Force} \times \text{Time} = \text{Mass} \times \text{velocity change}$$

$$\text{velocity change} = \frac{\text{Force} \times \text{Time}}{\text{Mass}} \tag{8.24}$$

In this case the velocity change due to this incremental impulse will be $\frac{Ff(\tau)d\tau}{M}$.

Recalling the solution for free vibration derived above (Equation 8.7)

$$x(t) = \frac{\dot{x}_o}{\omega} \sin \omega t + x_o \cos \omega t \tag{8.25}$$

If $\dot{x}_o$ is the initial velocity at the start of the increment of time and x_o is taken as zero (because there is no initial displacement corresponding to the effect of this impulse) then Δx, the displacement at time t due to the load applied during $d\tau$ is given by

$$\Delta x = \frac{Ff(\tau)d\tau}{M\omega} \sin \omega(t - \tau) \tag{8.26}$$

With a linear elastic system as here, superposition can be used to calculate the total displacement at t by summing of all impulse effects between $t = 0$ and t. Therefore we can write

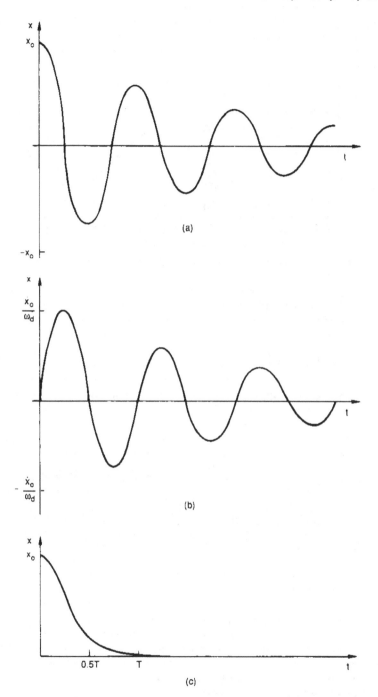

Figure 8.4 *Free vibration with viscous damping; (a) produced by initial displacement; (b) produced by initial velocity; (c) produced by initial displacement with critical damping (after Ref. 5)*

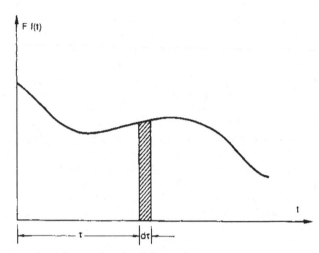

Figure 8.5   *Generalised forcing function (after Ref. 5)*

$$x(t) = \int_0^t \frac{Ff(\tau)}{M\omega} \sin \omega(t - \tau) \, d\tau$$

$$= x_{st}\omega \int_0^t f(\tau) \sin \omega(t - \tau) \, d\tau \tag{8.27}$$

where x_{st} is the static deflection due to F which is equal to F/K and which may be written as

$$x_{st} = \frac{F}{\omega^2 M} \tag{8.28}$$

Equation 8.27 is called the convolution or superposition integral and gives the solution for forced vibration of an SDOF undamped elastic structure for an arbitrary forcing function if initial displacement and velocity are zero. Complete generality is obtained by superimposing the effects of initial displacement and velocity as already evaluated producing the solution for displacement at time $t, x(t)$ as

$$x(t) = x_o \cos \omega t + \frac{\dot{x}_o}{\omega} \sin \omega t + x_{st}\omega \int_0^t f(\tau) \sin \omega(t - \tau) \, d\tau \tag{8.29}$$

In order to make this solution more specific to the case of a blast loaded structure we will now assess response to an idealised blast wave which can be represented by a forcing function in the shape of a triangular load pulse as shown in Figure 8.6 where t_d is the positive phase duration of the blast wave which produces a peak force on the structure F.

The equation of this forcing function is

$$F(t) = F\left(1 - \frac{t}{t_d}\right) \tag{8.30}$$

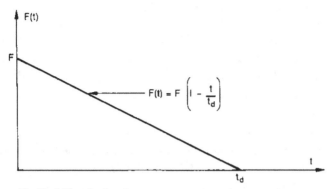

Figure 8.6 *Idealised blast load pulse*

Response can be assessed in two stages. The first is for times less than the positive phase duration of the blast. Here initial velocity and displacement are zero and $f(\tau)$ in Equation 8.29 is given by

$$f(\tau) = 1 - \frac{\tau}{t_d} \tag{8.31}$$

If these conditions are substituted into Equation 8.29 above we obtain

$$x(t) = x_{st}\omega \int_0^t f(\tau) \sin \omega(t - \tau) \, d\tau \tag{8.32}$$

which can be rewritten and integrated as set out below.

$$x(t) = \frac{F}{K}\omega \int_0^t \left(1 - \frac{\tau}{t_d}\right) \sin \omega(t - \tau) \, d\tau \tag{8.33}$$

$$= \frac{F}{K}\omega \left[\int_0^t \sin \omega(t - \tau) \, d\tau - \int_0^t \frac{\tau}{t_d} \sin \omega(t - \tau) \, d\tau \right]$$

$$= \frac{F}{K}\omega \left[\frac{1}{\omega} \cos \omega(t - \tau) - \frac{\tau}{t_d\omega} \cos \omega(t - \tau) + \int_0^t \frac{1}{t_d\omega} \cos \omega(t - \tau) \, d\tau \right]_0^t$$

$$= \frac{F}{K}\omega \left[\frac{1}{\omega} \cos \omega(t - \tau) - \frac{\tau}{t_d\omega} \cos \omega(t - \tau) + \frac{1}{t_d\omega^2} \sin \omega(t - \tau) \right]_0^t$$

$$= \frac{F}{K}\omega \left[\frac{1}{\omega} - \frac{t}{t_d\omega} - \frac{1}{\omega} \cos \omega t + \frac{1}{t_d\omega^2} \sin \omega t \right]$$

$$= \frac{F}{K} \left[1 - \frac{t}{t_d} - \cos \omega t + \frac{1}{t_d\omega} \sin \omega t \right] \tag{8.34}$$

Equation 8.34 can be written finally as

$$x(t) = \frac{F}{K}(1 - \cos \omega t) + \frac{F}{Kt_d}\left(\frac{\sin \omega t}{\omega} - t\right) \tag{8.35}$$

A dynamic load factor (DLF) or dynamic increase factor (DIF) can be defined as

$$DLF = \frac{x_{dyn}}{x_{st}} \tag{8.36}$$

where x_{dyn} is the (dynamic) displacement produced by the blast load. In this case, recalling that $x_{st} = F/K$, the DLF is given by

$$DLF = 1 - \cos \omega t + \frac{\sin \omega t}{\omega t_d} - \frac{t}{t_d} \tag{8.37}$$

The second stage of response is for times in excess of the positive phase duration of the blast load. Initial displacement and velocity are now given by

$$x_o = \frac{F}{K}\left(\frac{\sin \omega t_d}{\omega t_d} - \cos \omega t_d\right)$$

$$\dot{x}_o = \frac{F}{K}\left(\omega \sin \omega_d t + \frac{\cos \omega t_d}{t_d} - \frac{1}{t_d}\right) \tag{8.38}$$

$$f(\tau) = 0$$

Substituting these values into the general solution above and replacing t by $(t - t_d)$ and simplifying yields

$$x(t) = \frac{F}{K\omega t_d}[\sin \omega t_d - \sin \omega(t - t_d)] - \frac{F}{K}\cos \omega t$$

$$DLF = \frac{1}{\omega t_d}[\sin \omega t_d - \sin \omega(t - t_d)] - \cos \omega t \tag{8.39}$$

Typical responses for this idealised blast loading are compared in Figure 8.7 for two values of t_d/T where T is the natural response period of the structure $(= 2\pi/\omega)$. When the ratio of positive phase duration to natural response period is large, the blast wave loading is still present while the structure undergoes a number of vibrations. This situation will be referred to below as 'quasi-static' or 'pressure' loading and could be associated with the response of a structure to a long duration load such as might be produced by a large yield nuclear device or the pulse produced by a vapour cloud explosion. When the ratio is small this implies that the blast wave loading has finished and the structure has not completed even a single cycle of response. This situation will be referred to below as 'impulsive' loading.

In Figure 8.8 the variation of maximum dynamic load factor, DLF_{max} with t_d/T is shown while in Figure 8.9 the time taken by the structure to reach maximum response (t_m) is presented.

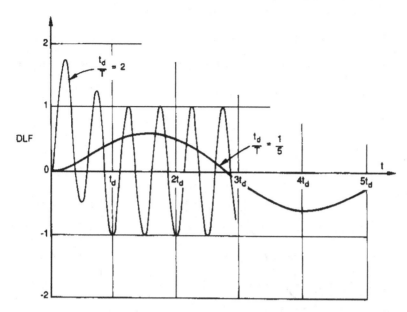

Figure 8.7 *Response of SDOF system to idealise blast loading for different* t_d/T *ratios (after Ref. 5)*

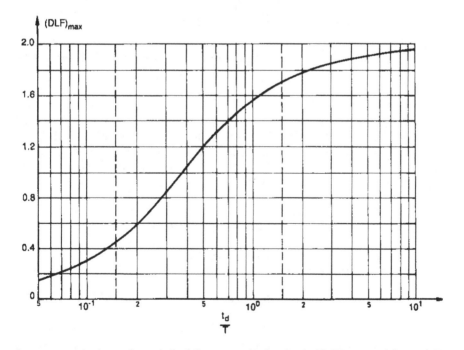

Figure 8.8 *Maximum dynamic load factor vs* t_d/T *for elastic SDOF system (after Ref. 5)*

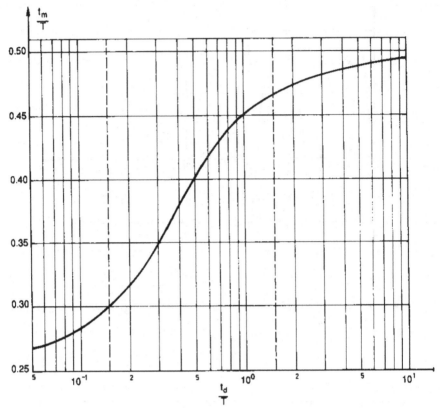

Figure 8.9   *Time to reach maximum displacement vs t_d/T for elastic SDOF system (after Ref. 5)*

Inspection of Figure 8.8 shows that DLF_{max} tends to a value of 2 as t_d/T becomes large: the effect of the decay in blast force is negligible in the time required for the response to reach the first peak. Clearly knowledge of the t_d/T ratio is vital in assessing structural response and further consideration of this ratio will be presented later.

8.7   Forced vibration of damped system (*c* finite)

A generalised solution for the forced vibration of a damped system may be obtained in a similar manner to that used for an undamped system as described above.

In this case the incremental displacement Δx due to an elemental impulse becomes

$$\Delta x = \frac{Ff(\tau)d\tau}{M\omega_d}e^{-\zeta\omega(t-\tau)}\sin\,\omega_d(t-\tau) \qquad (8.40)$$

It is worth noting, of course, that for lightly damped systems in which, say, ζ is about 0.1, then ω_d is to a good approximation equal to ω.

The complete analysis, which is carried out in the manner described above for undamped systems, will not be detailed here but is available in Reference 1. The result of such analysis gives the displacement x at time t as

$$x(t) = e^{-\zeta \omega t} \left[\frac{\dot{x}_0 + \omega \zeta x_0}{\omega_d} \sin \omega_d t + x_0 \cos \omega_d t \right]$$

$$+ x_{st} \frac{\omega^2}{\omega_d} \int_0^t f(\tau) e^{-\zeta \omega (t - \tau)} \sin \omega_d (t - \tau)\, d\tau \tag{8.41}$$

This solution is identical to the solution of an undamped system if $\zeta = 0$.

8.8 Resonance of a damped and undamped structure

In order to illustrate the phenomenon of resonance, consider a mass supported on an elastic spring loaded by sinusoidal forcing function. A damping component of resistance is also included as shown in Figure 8.10.

The equation of motion, derived by considering the free body diagram, is

$$M\ddot{x} + c\dot{x} + Kx = F \sin \omega_f t \tag{8.42}$$

Then by rearranging and substituting ω and ζ the equation becomes

$$\ddot{x} + 2\zeta \omega \dot{x} + \omega^2 x = \frac{F}{M} \sin \omega_f t \tag{8.43}$$

which has a solution which can be written as

$$x = x_{max} \sin [\omega_f t - \varphi] \tag{8.44}$$

where φ is the phase of the displacement with respect to the exciting force. If this solution is substituted into equation 8.43 and noting that the term $\sin \omega_f t$ can be expanded as shown

$$\sin \omega_f t = \sin [(\omega_f t - \varphi) + \varphi] = \sin (\omega_f t - \varphi) \cos \varphi + \cos (\omega_f t - \varphi) \sin \varphi \tag{8.45}$$

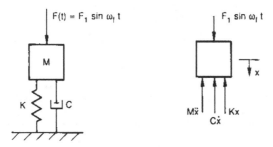

Figure 8.10 *SDOF system with sinusoidal forcing function*

substitution and simplification results in

$$\left[-\omega_f^2 x_{max} + \omega^2 x_{max} - \frac{F}{M} \cos \varphi\right] \sin (\omega_f t - \varphi)$$

$$+ \left[2\omega\zeta x_{max} - \frac{F}{M} \sin \varphi\right] \cos (\omega_f t - \varphi) = 0 \qquad (8.46)$$

which can only be generally true if coefficients in the square brackets are zero. Therefore we obtain from the first term

$$x_{max} = \frac{F}{M(\omega^2 - \omega_f^2)} \cos \varphi \qquad (8.47)$$

and from the second

$$x_{max} = \frac{F}{2\omega\zeta\omega_f M} \sin \varphi \qquad (8.48)$$

The parameter ζ can be eliminated by obtaining $\cos^2\varphi + \sin^2\varphi = 1$ to give

$$x_{max} = \frac{F/K}{\sqrt{[1 - (\omega_f/\omega)^2]^2 + [2\zeta\omega_f/\omega]^2}} \qquad (8.49)$$

and

$$\tan \varphi = \frac{2\zeta\frac{\omega_f}{\omega}}{1 - \left(\frac{\omega_f}{\omega}\right)^2} \qquad (8.50)$$

The dynamic load factor in this case is

$$DLF = \frac{x_{max}}{(F/K)} = \frac{1}{\sqrt{[1 - (\omega_f/\omega)^2]^2 + [2\zeta\omega_f/\omega]^2}} \qquad (8.51)$$

In blast loading response assessment it is often the case that maximum displacement is of principal interest. Since this occurs during the first cycle of oscillation when damping is relatively unimportant (and its omission leads to an overestimate of response and ultimately to a conservative design) c and hence ζ are set to zero. Results for both damped and undamped responses are plotted in Figure 8.11 as DLF vs ω_f/ω.

When the structure is undamped and the frequency of the loading function matches the natural frequency of the structure, resonance occurs and the theoretical amplitude of vibration of the structure is infinitely large.

Example 8.1
A structure has been idealised as a lumped-mass SDOF system of mass M with a resistance characterised by an elastic spring of stiffness K. How will the structure respond if it is loaded by a long duration blast wave from a large-scale nuclear device which may be approximated to a suddenly applied force of F which lasts for an infinitely long time? This situation could be described as quasi-static loading. The situation is illustrated in Figure 8.12 below.

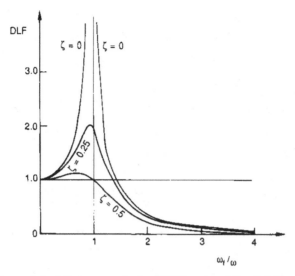

Figure 8.11 *Dynamic load factor vs ω_f/ω for SDOF system (after Ref. 1)*

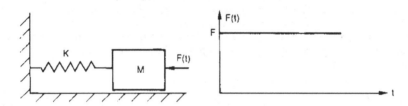

Figure 8.12 *Idealised SDOF system for Example 8.1*

The general solution to this problem is given by Equation 8.29 above. In this case assume that the structure is initially at rest and undisplaced from its equilibrium position. Given that the forcing function is F, Equation 8.29 reduces to

$$x(t) = x_{st}\omega \int_0^t \sin \omega(t - \tau) \, d\tau \tag{8.52}$$

which can be integrated to give the displacement at time t as

$$x(t) = x_{st}(1 - \cos \omega t) \tag{8.53}$$

This response can be represented graphically in Figure 8.13 where it will be seen that under the action of this loading the structure will oscillate about a mean displacement equal to the displacement if the load F were applied statically. The maximum value of the *DLF* is, of course, 2.

If this were a real structure, the likelihood would be that it would undergo yield at a particular displacement. After this point the nature of the resistance that the structure offers to the applied loading changes and the equation to be solved will alter and new initial conditions will need to be specified. Depending on the magnitude of the blast load and the maximum value of resistance that the structure develops, a third form of the equation would need to be solved to account for the unloading behaviour

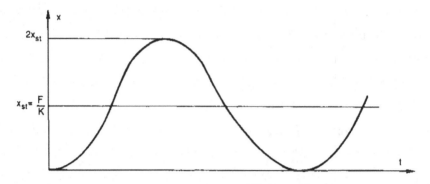

Figure 8.13 *Displacement-time history for SDOF system of Example 8.1*

of the now plastically deforming structure. This aspect will be considered in more detail below.

Example 8.2

As a contrast to the situation of Example 8.1, the same structure again, initially undisplaced from its equilibrium position, is loaded by a blast wave produced by the detonation of a quantity of high explosive at close range. The overpressure loading is high but of very short duration. As a consequence the structure may be considered as having been loaded by an impulse I imparting an initial velocity $\dot{x}_0$ to the structure given by

$$\dot{x}_0 = \frac{I}{M} \tag{8.54}$$

In this case equation 8.29 reduces to

$$x(t) = \frac{I}{M\omega} \sin \omega t \tag{8.55}$$

and the response, which is one of essentially free vibration, is as illustrated in Figure 8.14.

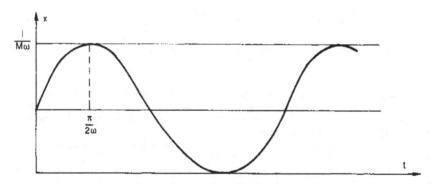

Figure 8.14 *Displacement-time history for SDOF system of Example 8.2*

In this case oscillation is about a zero mean amplitude with maximum displacement equal to $I/M\omega$.

8.9 References

[1] Thomson W.T., *Theory of Vibrations with Applications*. (3rd edition) Unwin Hyman, London (1988)
[2] Irvine H.M., *Structural Dynamics for the Practising Engineer*. Allen and Unwin, London (1986)
[3] Berg G.V., *Elements of Structural Dynamics*. Prentice-Hall, New Jersey (1989)
[4] Paz M., *Structural Dynamics: Theory and Computation*. Van Nostrand-Reinhold, New York (1980)
[5] Biggs J.M., *Introduction to Structural Dynamics*. McGraw-Hill, New York (1964)
[6] TM5-1300 Design of Structures to Resist the Effects of Accidental Explosions. *US Department of the Army Technical Manual* (1991)

Symbols

c	damping coefficient
D	operator for solution of second order differential equations
DIF	dynamic increase factor
DLF	dynamic load factor
$f(t)$	time-varying loading
$F(t)$	forcing function
I	impulse
K	structure elastic resistance
M	structure mass
N	normal reaction force
t	time
t_d	positive phase duration of idealised triangular blast load
t_m	time to reach maximum displacement
T	natural response period of structure
$x(t)$	time-dependent displacement
x_o	initial displacement
$\dot{x}_o$	initial velocity
x_{dyn}	maximum dynamic displacement
x_{st}	maximum static displacement
Δx	displacement at time t due to impulse of duration $d\tau$
ζ	damping factor
μ	coefficient of friction
$d\tau$	element of time during which an incremental impulse is delivered
ω	natural circular frequency of vibration
ω_d	natural circular frequency of damped structure
ω_f	circular frequency of forcing function
φ	phase lag angle

9 Structural response: pressure-impulse diagrams

9.1 Introduction

In Chapter 8 the principles of structural dynamics were introduced with emphasis placed on the response of single degree of freedom systems to forcing functions that represented idealised blast loading histories. In this chapter these ideas will be developed towards assessment of the response of specific structural elements and then to convert these flexing, distributed mass structures back to single degree of freedom lumped mass 'equivalent' systems.

In assessing the behaviour of a blast loaded structure it is often the case that the calculation of *final states* is the principal requirement for a designer. He needs to know what is the maximum displacement of a structure that is acceptable given a specified form of blast loading, rather than a detailed knowledge of the displacement-time history of a structure. This approach will be adopted here and will be seen to lead to a *failure* criterion for specific structures by the development of pressure-impulse diagrams.

To establish the principles of this analysis, consideration of the problem will commence by taking the response assessment for a single degree of freedom (SDOF) elastic structure presented in Chapter 8 further. Finally, the chapter presents damage criteria for specific target structures, in this case brick-built houses and similar structures, and also for the human body, by means of empirically developed pressure-impulse diagrams and similar injury prediction graphs.

9.2 Rigid target elastically supported

Consider a target structure which has been idealised as an SDOF elastic structure and which is to be subjected to a blast load idealised as a triangular pulse delivering a peak force F to the target. The positive phase duration of the blast load is t_d. The situation is illustrated in Figure 9.1.

The load pulse is described by Equation 8.30 reproduced here

$$F(t) = F\left(1 - \frac{t}{t_d}\right) \tag{9.1}$$

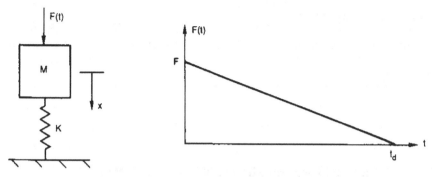

Figure 9.1 *Single degree of freedom system loaded by triangular blast load pulse*

This blast load will deliver an impulse I to the target structure given by the equation

$$I = \frac{1}{2}Ft_d \tag{9.2}$$

where I is the area beneath the load function for $0 < t < t_d$. The equation of motion for this structure is

$$M\ddot{x} + Kx = F\left(1 - \frac{t}{t_d}\right) \tag{9.3}$$

This equation has already been solved in detail in Chapter 8. If we confine the problem to the response for times less than the positive phase duration of the blast wave the solution is given by Equation 8.35 which is rewritten below.

$$x(t) = \frac{F}{K}(1 - \cos \omega t) + \frac{F}{Kt_d}\left(\frac{\sin \omega t}{\omega} - t\right) \tag{9.4}$$

If we limit ourselves to the worst case of response then we need the maximum dynamic structure displacement x_{max} which will occur when the velocity of the structure $\dot{x}$ is zero. Therefore we can differentiate Equation 9.4 and obtain

$$0 = \omega \sin (\omega t_m) + \frac{1}{t_d} \cos (\omega t_m) - \frac{1}{t_d} \tag{9.5}$$

In this equation t_m is the time at which the displacement reaches x_{max}. Equation 9.5 may be solved by trial and error or otherwise to obtain a relationship of the general form

$$\omega t_m = f(\omega t_d) \tag{9.6}$$

From this general relationship it is clear that a similar form of equation can be obtained for maximum dynamic displacement

$$\frac{x_{max}}{F/K} = \psi(\omega t_d) = \psi'\left(\frac{t_d}{T}\right) \tag{9.7}$$

where ψ and ψ' are functions of ωt_d and t_d/T respectively. This solution re-emphasises the finding of Chapter 8 that there is a strong relationship between both the structure natural frequency and hence natural response period T and the positive phase duration of the blast wave as characterised by t_d. To proceed further it is essential to consider now the relative magnitudes of these quantities.

9.3   Positive phase duration and natural period

9.3.1   Positive phase long compared with natural period

Let consideration first be given to the situation where the blast load duration is much larger than the natural period of vibration of the structure. Given the relationship between T and ω and between ω and structure mass M and stiffness K, we can write for this case

$$t_d \gg T = 2\pi/\omega\left(\propto 1/\omega = (M/K)^{1/2}\right) \tag{9.8}$$

In the limit the load may be considered as remaining constant while the structure attains its maximum deflection. For example, this could be the case for a structure loaded by a blast wave derived from a nuclear device at medium to long range. In this case the maximum displacement x_{max} is solely a function of the peak blast load F and the stiffness K and does not involve either the positive phase duration or the mass of the structure. The situation can be represented graphically in Figure 9.2 which shows the variation of both blast load and structure resistance $R(t)$ with time. The structure is seen to have reached its maximum displacement before the blast load has undergone any significant decay. Such loading is referred to as quasi-static or pressure loading.

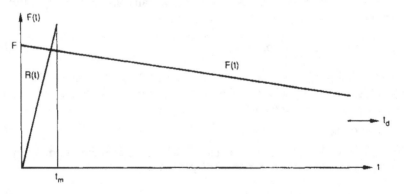

Figure 9.2   *Quasi-static loading: response time compared with load duration*

9.3.2 Positive phase short compared with natural period

Now consider the situation in which the positive phase duration of the blast t_d is much shorter than the natural response period of the target structure. In this case the load has finished acting before the structure has had time to respond significantly – most deformation occurs at times greater than t_d.

In this case we can write

$$x_{max} = f(I, K, M) \tag{9.9}$$

This situation can be represented graphically in Figure 9.3

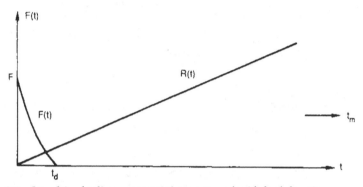

Figure 9.3 *Impulsive loading: response time compared with load duration*

Inspection of this graph indicates that the blast load pulse has fallen to zero before any significant displacement occurs. In the limit, the blast load will be over before the structure has moved at all. The structure in this situation is described as being subjected to impulsive loading.

9.3.3 Positive phase duration and natural period similar

In this case, with t_d and T approximately the same, the assessment of response in this regime is more complex, possibly requiring complete solution of the equation of motion of the structure, though it is often possible to make reasonable approximations of response by using results obtained for impulsive or quasi-static loading. This dynamic or pressure-time regime can be represented graphically as shown in Figure 9.4.

Methods of solution of the equation of motion, particularly for complex resistance functions, will be discussed in Chapter 11.

These three regimes can be summarised in terms of the product of natural frequency and positive phase duration which is proportional to the ratio of T to t_d as indicated in Equation 9.10.

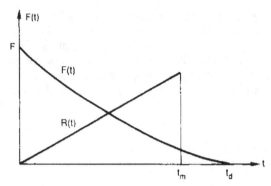

Figure 9.4   *Dynamic loading: response time compared with load duration*

$$0.4 > \omega t_d \quad \left[\propto \frac{t_d(\text{short})}{T(\text{long})} \right] \quad (\text{Impulsive})$$

$$40 < \omega t_d \left[\propto \frac{t_d(\text{long})}{T(\text{short})} \right] \quad (\text{Quasi} - \text{static})$$

$$0.4 < \omega t_d < 40 \left[\frac{t_d}{T} \approx 1 \right] \quad (\text{Dynamic}) \tag{9.10}$$

Alternatively the regimes can be represented in terms of the time taken to reach maximum displacement t_m compared with positive phase duration as indicated in Figure 9.5 which has been adapted from the 1969 edition of Reference 11.

9.4   Evaluation of the limits of response

In this section the principles for evaluating the response of structures loaded quasi-statically and impulsively will be described with reference to the simple structure of Figure 9.1 by considering the energy of the system.

In the case of quasi-static loading the load pulse can be idealised as shown in Figure 9.6a while the structure resistance can be represented by the graph of Figure 9.6b.

The basic principle of the analysis is to specify that the work done on structure is all converted to strain energy as the structure deforms. The work done by the load as it causes a displacement x_{max} is WD given by

$$WD = Fx_{\text{max}} \tag{9.11}$$

The strain energy acquired by the structure, U, is the area beneath the resistance displacement graph given by

$$U = \frac{1}{2} K x_{\text{max}}^2 \tag{9.12}$$

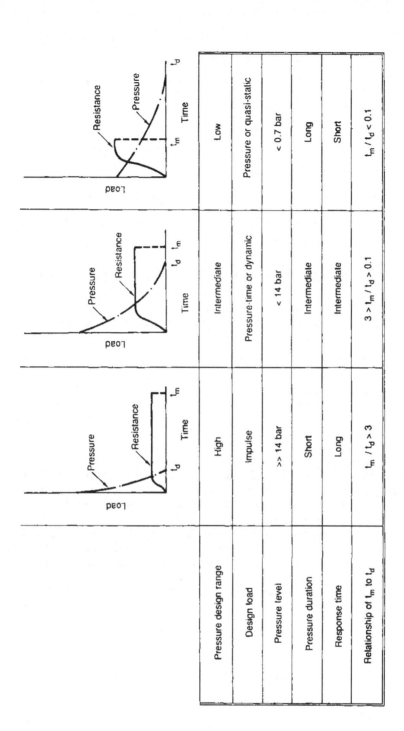

	High	Intermediate	Low
Pressure design range			
Design load	Impulse	Pressure-time or dynamic	Pressure or quasi-static
Pressure level	>> 14 bar	< 14 bar	< 0.7 bar
Pressure duration	Short	Intermediate	Long
Response time	Long	Intermediate	Short
Relationship of t_m to t_d	$t_m / t_d > 3$	$3 > t_m / t_d > 0.1$	$t_m / t_d < 0.1$

Figure 9.5 *Load regime summary (after Ref. 9)*

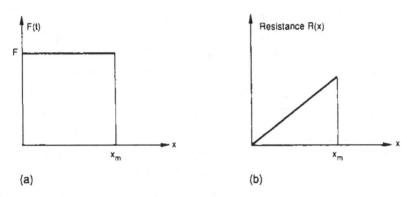

Figure 9.6   *(a) Idealised quasi-static load pulse; (b) elastic resistance vs displacement*

If *WD* and *U* are equated and the resulting equation rearranged then

$$\frac{x_{max}}{(F/K)} = 2 \tag{9.13}$$

which can be written as

$$\frac{x_{max}}{x_{st}} = 2 \tag{9.13}$$

where x_{st} is the static displacement that would result if the force *F* were applied statically. Equation 9.13 is, of course, the dynamic load factor (*DLF*) introduced in Chapter 8. In this limiting case it takes on the maximum value of 2 and represents the upper bound of response and is referred to as the quasi-static asymptote.

To analyse the response of the structure to an impulsive load the principle involved is to acknowledge that when an impulse is delivered to a structure it produces an instantaneous velocity change. As a consequence the structure gains kinetic energy which is converted to strain energy.

The impulse causes an initially stationary structure to acquire a velocity $\dot{x}_0$ given by

$$\dot{x}_0 = \frac{I}{M} \tag{9.14}$$

Hence the pulse delivers kinetic energy *KE* to the structure given by

$$KE = \frac{1}{2}M\dot{x}_0^2 = \frac{I^2}{2M} \tag{9.15}$$

As the structure displaces it will acquire the same strain energy *U* as before as it displaces by x_{max}. Thus if *KE* and *U* are equated and the resulting equation rearranged to make x_{max} the subject, division by *F/K* leads to the following

$$\frac{I^2}{2M} = \tfrac{1}{2}Kx_{max}^2$$

$$\frac{x_{max}}{F/K} = \frac{I}{\sqrt{KM}(F/K)} = \frac{\tfrac{1}{2}Ft_d}{\sqrt{KM}(F/K)} = \tfrac{1}{2}\omega t_d \qquad (9.16)$$

which is the equation of the impulsive asymptote of response.

These two asymptotes can be drawn on a response curve with an abscissa of ωt_d and an ordinate of $\frac{x_{max}}{F/K}$. The actual response of the structure can then be sketched (without further analysis) relative to the asymptotes as shown in Figure 9.7

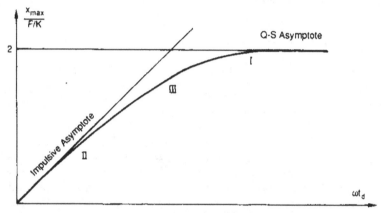

Figure 9.7 *Response of elastic SDOF system for all load regimes*

The three regimes of quasi-static, impulsive and dynamic response are identified on the resulting graph as regions I, II and III respectively.

9.5 Pressure-impulse or iso-damage curves

Although the graph of Figure 9.7 is useful, it is possible, with relatively little extra work, to convert this representation of response to a pressure-impulse (P-I) or iso-damage curve which allows the load-impulse combination that will cause a specified level of damage to be assessed very readily. The bounds on behaviour of a target structure are characterised by a pressure (though in the specific example here force is used) and an impulse (which can be a total impulse I (as here) or a specific impulse i_s or i_r).

From the previous analysis, the response curve had an ordinate which has the value two for quasi-static loading. Thus

$$\frac{x_{max}}{F/K} = \frac{Kx_{max}}{F} = 2 \qquad (9.17)$$

which can be written

$$\frac{Kx_{max}}{2F} = 1 \tag{9.18}$$

If this is now inverted we obtain

$$\frac{2F}{Kx_{max}} \left[\propto \frac{\text{Maximum Load}}{\text{Maximum Resistance}} \right] = 1 \tag{9.19}$$

which is the equation of a modified quasi-static asymptote plotted on a graph with ordinate $2F/Kx_{max}$. The abscissa of Figure 9.7 is ωt_d. If this is multiplied by the inverse of the ordinate of Figure 9.7, and recalling that $\omega = \sqrt{K/M}$ then we obtain

$$\frac{Ft_d}{Kx_{max}} \sqrt{\frac{K}{M}} = \frac{2I}{x_{max}\sqrt{KM}} \tag{9.20}$$

which is a non-dimensionalised impulse. Noting that the impulsive asymptote of Figure 9.7 is given by Equation 9.16, using an abscissa $\frac{I}{x_{max}\sqrt{KM}}$ means that the new impulsive asymptote is given by $\frac{I}{x_{max}\sqrt{KM}} = 1$. Now the response curve can be replotted using axes $\frac{2F}{Kx_{max}}, \frac{I}{x_{max}\sqrt{KM}}$ to obtain Figure 9.8.

The form of this graph allows easier assessment of response to a specified load. Once a maximum displacement is defined (i.e. a damage criterion has been specified) this curve then indicates the combinations of load and impulse that will cause failure. Combinations of pressure and impulse that fall to the left of and below the curve will not induce failure while those to

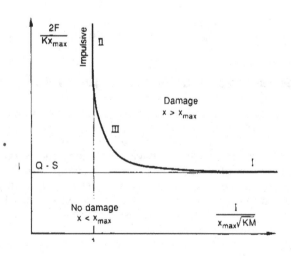

Figure 9.8 *Pressure-impulse diagram for elastic SDOF (after Ref. 4)*

the right and above the graph will produce damage in excess of the allowable limit.

Other structural models can be considered in the same way. Consider a structure of mass M idealised as a rigid-plastic structure which undergoes yield when it reaches a resistance force R as shown in Figure 9.9.

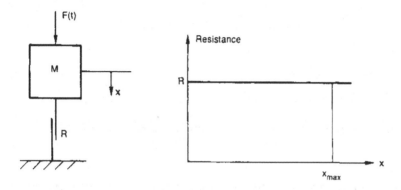

Figure 9.9 *Rigid-plastic structure loaded by blast wave (after Ref. 4)*

Analysis proceeds as before to establish the upper and lower bounds of response, the quasi-static and impulsive asymptotes respectively. The work done as the (constant) blast force F moves through a distance x_{max} is equated to the strain energy acquired by the structure. Thus

$$Fx_{max} = Rx_{max}$$

$$\therefore \quad \frac{F}{R} = 1 \tag{9.21}$$

This idealisation, of course, means that the quasi-static asymptote is also the zero deflection asymptote and, for force to resistance ratios that just exceed unity, large displacements can occur.

By equating the kinetic energy delivered by the blast load impulse I to strain energy the impulsive asymptote is obtained. Thus

$$\frac{I^2}{2M} = Rx_{max}$$

$$\therefore \quad \frac{I^2}{x_{max}MR} = 2$$

$$\therefore \quad \frac{I}{\sqrt{x_{max}MR}} = \sqrt{2} \tag{9.21}$$

Using axes $\frac{I}{\sqrt{x_{max}MR}}, \frac{F}{R}$ the iso-damage curve is as shown in Figure 9.10.

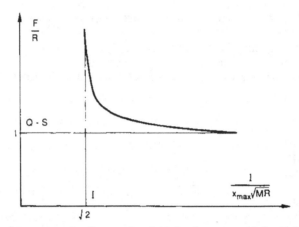

Figure 9.10   *Pressure-impulse diagram for rigid-plastic structure (after Ref. 4)*

Example 9.1

As a further example consider a structure idealised as an SDOF system of mass M loaded by a blast wave producing a peak load F which is triangular in shape and of positive phase duration t_d. The structure exhibits elastic-plastic resistance with elastic stiffness K. At yield the displacement is x_{el} after which the structure can displace to x_{max}. The system is illustrated in Figure 9.11.

It is fairly straightforward to develop pressure-impulse diagrams for different levels of damage, that is, increasing values of x_{max}. Thus, equating work done and strain energy and kinetic energy and strain energy allows evaluation of suitable pressure-impulse diagram axes and the location of the asymptotes. For quasi-static response

$$Fx_{max} = \tfrac{1}{2}x_{el}R_{max} + R_{max}(x_{max} - x_{el}) \tag{9.21}$$

Since structure resistance R in the elastic region is Kx, at yield $R_{max} = Kx_{el}$.

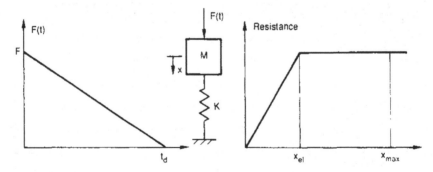

Figure 9.11   *Elastic-plastic SDOF system loaded by triangular blast load pulse*

If this relationship is substituted into equation 9.21 then, with some rearrangement

$$Fx_{max} = Kx_{el}^2 \left(\frac{x_{max}}{x_{el}} - \frac{1}{2} \right) \tag{9.22}$$

If x_{max} is written as nx_{el} where $n > 1$ then the quasi-static asymptote is given by

$$\frac{F}{Kx_{max}} = \frac{2n-1}{2n^2} \tag{9.23}$$

For the impulse regime response we obtain

$$\frac{I^2}{2M} = Kx_{max}^2 \left(\frac{2n-1}{2n^2} \right) \tag{9.24}$$

from which the impulsive asymptote can be written as

$$\frac{I}{\sqrt{KM}x_{max}} = \frac{\sqrt{2n-1}}{n} \tag{9.25}$$

To allow more direct assessment of force and impulse to cause different levels of damage the expressions of Equations 9.23 and 9.25 can be simplified by choosing particular values of mass, stiffness and maximum elastic displacement. If each of these quantities is set equal to unity using appropriate units, the axes of the resulting iso-damage curves become F and I given by

$$F = \frac{2n-1}{2n}$$
$$I = \sqrt{2n-1} \tag{9.26}$$

To be even more specific, if the levels of damage are selected as twice, three and four times the maximum elastic displacement (i.e. $n = 2, 3$ and 4), a series of iso-damage curves for this structure can be drawn relative to the resulting three sets of asymptotes as shown in Figure 9.12.

It will be seen that a higher level of force and/or impulse is required to produce more severe damage.

Whereas the foregoing relates to analytically derived pressure-impulse diagrams it is possible to derive iso-damage curves from experimental evidence or real-life events. Iso-damage curves have been derived from a study of houses damaged by bombs dropped on the United Kingdom in the Second World War.[1] The results of such investigations are used as part of the evaluation of safe-stand-off distances for explosive testing in the United Kingdom. In this instance the axes of the curves are simply side-on peak overpressure p_s and side-on specific impulse i_s as shown in Figure 9.13.

The levels of damage are less precise than those obtained by analysis above. As quoted in Reference 1, they are:

Category A – Almost complete demolition.
Category B – Such severe damage as to necessitate demolition – 50–75% of external brickwork destroyed or unsafe.

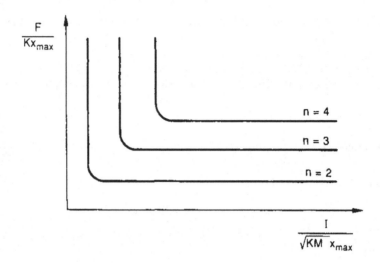

Figure 9.12 *Pressure-impulse diagrams for elastic-plastic SDOF system for different damage levels*

Category C_b – Damage rendering house temporarily uninhabitable – partial collapse of roof and one or two external walls. Load-bearing partitions severely damaged and requiring replacement.

Category C_a – Relatively minor structural damage yet sufficient to make house temporarily uninhabitable. Partitions and joinery wrenched from fixings.

Category D – Damage calling for urgent repair but not so as to make house uninhabitable. Damage to ceilings and tiling. More than 10% of glazing broken.

The curves of Figure 9.13 can also be used with reasonable confidence to predict the damage to other structures such as small office buildings and light-framed factories.

The use of pressure-impulse diagrams in conjunction with blast parameter vs scaled distance graphs allows the development of equations to describe specific damage levels. An example for the curves above is of the general form

$$R = \frac{K'W^{1/3}}{[1 + (3175/W)^2]^{1/6}} \tag{9.27}$$

where R is range in metres, W is weapon yield in kilogrammes of TNT and K' is an empirical constant. The value of K' in the equation above giving the radius of Category B damage R_B is 5.6. Radii for Categories A and C_b are given approximately by 0.675 R_B and 1.74 R_B respectively.

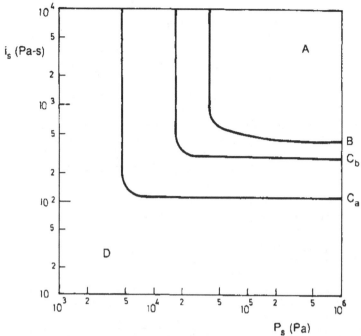

Figure 9.13 *Pressure-impulse diagrams for damage to houses and small buildings (after Ref. 4)*

Development of such equations in turn leads to the creation of range-charge weight (R-W) overlays which can be used in conjunction with P-I diagrams to assess the damage potential of a specific threat to a target. The general form of such an overlay is shown in Figure 9.14.

This figure indicates that pressure p_1 (say) will be produced by a charge of mass W_1 at a range R_1 or by charge W_2 at range R_2 and so on as determined from evaluation of the scaled distance Z corresponding to p_1. Also, of course, at this Z value there is a unique value of $i_s / W^{1/3}$ so that each charge weight W has a corresponding value of i_s. This procedure can be repeated for different pressures and impulses. It is evident that the particular level of damage specified by the iso-damage curve of Figure 9.14 will be produced by 1 kg of TNT at a range of 0.25 m or by 2 kg of TNT at 0.75 m or by 3 kg at 1 m range.

Consider now a more specific situation relating to bomb damage to houses in the Second World War 2 shown in Figure 9.13. It is a relatively straight-forward procedure to superimpose an R-W overlay so that the effects of specific weapons at different ranges can be compared. Figure 9.15 presents such an overlay using information derived from References 2 and 3. GP250 and GP2000 are general purpose bombs of all-up weight 118 kg and 895 kg respectively containing explosive of TNT of equivalent weight 55 kg and 542 kg respectively. As well as these bombs 2.2 t TNT and 10 t TNT are considered, assumed to be detonated as hemispherical stacks on the ground. Inspection of the graph reveals, for instance, that the GP250

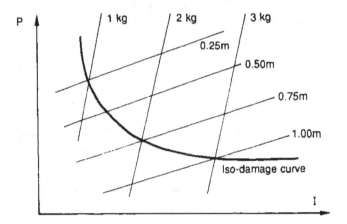

Figure 9.14 *Pressure-impulse diagram with range-charge weight overlay*

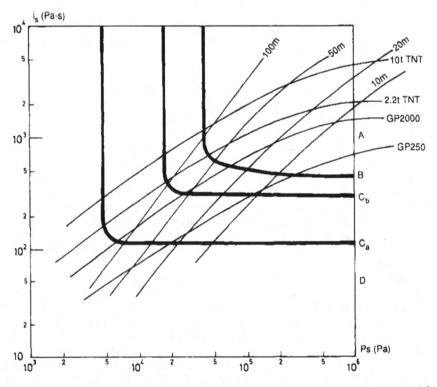

Figure 9.15 *Pressure-impulse diagram for damage to houses with range-charge weight overlay*

bomb at 10 m range is likely to cause Category C_b damage while at 20 m range damage is Category C_a. At 50 m range damage to dwellings from this bomb is likely to be relatively minor. On the other hand even at 100 m from the large-scale explosive stack demolition is likely to be total.

9.6 Pressure-impulse diagrams for gas and dust explosions

As described in Chapter 4, pressure-time histories for gas and dust explosions differ from those produced by condensed explosives in that for gas and dust events it is necessary to define explosion intensity not only in terms of peak overpressure and duration but also in terms of maximum pressure rise rate. A typical pressure-time curve for an unconfined gas or dust explosion is shown in Figure 9.16.

Baker *et al.*[4] propose that the shape of the curve can be given approximately by the equation

$$p(t) = p_m \left[\frac{t}{t_r} - \frac{1}{2\pi} \sin\left(\frac{2\pi t}{t_r}\right) \right] \tag{9.23}$$

for times less than the time to reach peak overpressure, t_r, and by

$$p(t) = p_m \left[1 - \left(\frac{t - t_r}{t_d - t_r}\right) \right] \exp\left\{ -\left(\frac{t - t_r}{t_d - t_r}\right) \right\} \tag{9.24}$$

for time between t_r and t_d. At $t = t_r/2$, dp/dt is maximum and from Equation 9.23 this is seen to be

$$\frac{dp}{dt}\bigg|_{max} = \frac{2p_m}{t_r} \tag{9.25}$$

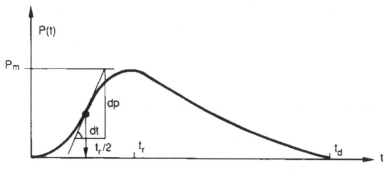

Figure 9.16 *Pressure-time history of confined gas or dust explosion*

In Figure 9.17 the response of an SDOF elastic structure to this loading is compared with that produced by a blast load from a condensed explosive.

Points to note on this figure are that the quasi-static asymptote for the gas or dust explosion is at 1.0 meaning that, because the load is applied relatively slowly to the structure, there is no dynamic increase factor and the loading may therefore be considered as the equivalent of a static load. The curve for the condensed explosive displays a dynamic load factor of 2.0. For $1.15 < \frac{I}{x_{max}\sqrt{KM}} < 5.5$ loading from a gas or dust event is more severe than for condensed explosive because of resonance between load rate and structure frequency. Finally, it is observed that if t_r/t_d varies, only small changes in the response curves are produced.

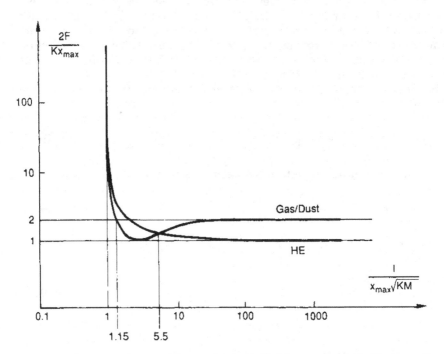

Figure 9.17   *Pressure-impulse diagrams for elastic SDOF system loaded by blast wave from condensed HE and gas or dust explosion (after Ref. 4)*

9.7   Human response to blast loading

In this section pressure-impulse diagrams to allow prediction of personnel injury from blast loading will be developed together with methods of assessing other forms of blast-related injury.

Most analyses of human response to blast loading assume that the subject is being loaded by free-field blast waves travelling over flat, level ground. However, this may not be the worst case of loading. Particularly, the proximity of a reflecting surface can cause load enhancement as shown in Figure 9.18a with the subject experiencing a more complex load environment than the subject in Figure 9.18b.

Of course, it is not only the direct effect blast wave overpressure that produces injury and three categories of blast-induced injury have been identified.

(a) *Primary* injury is due directly to blast wave overpressure and duration which can be combined to form specific impulse. Overpressures are induced in the body following arrival of the blast and the level of injury sustained depends on a person's size, gender and (possibly) age. The location of most severe injuries is where density differences between adjacent body tissues are greatest. Likely damage sites thus include the lungs which are prone to haemorrhage and oedema (which is the collection of fluid), the ears (particularly the middle ear) which can rupture, the larynx, trachea and the abdominal cavity.

(b)*Secondary* injury is due to impact by missiles (fragments from weapon casing, for example) created by explosive devices. Such missiles produce lacerations, penetration and blunt trauma (a severe form of bruising).

(c) *Tertiary* injury is due to displacement of the entire body which will be inevitably followed by high decelerative impact loading when, of course,

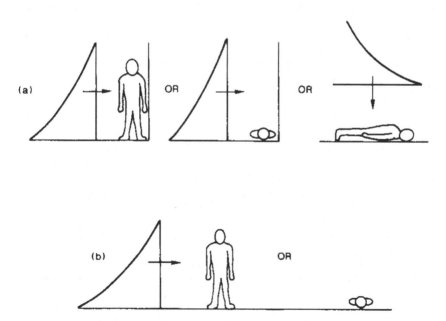

Figure 9.18 *(a) Subject near reflecting surfaces loaded by blast wave; (b) subject remote from reflecting surface loaded by blast wave (after Ref. 4)*

most damage occurs. Even when the person is wearing a good protective system, skull fracture is possible.

9.7.1   Lung damage

When a blast wave interacts with the torso two sorts of response occur: stress waves (see Chapter 6) are generated which interact with different components of the body, and the entire torso suffers gross compression. Which of these two elements produces the most damaging response is still uncertain.

Consider specifically the lungs which, being composed of air sacs called alveoli, are consequently less dense and also less stiff than the surrounding tissues. One mode of response to blast loading is for the alveoli to compress due to the implosion of the chest wall and the upward movement of the diaphragm. This process is very rapid and the body components cannot respond quickly enough to accommodate these imposed changes. The result is distortion of the lungs which suffer shearing. It is likely that this process is also allied with stress wave interactions between low density lung tissue and the more dense surrounding material. The result is tissue rupture which

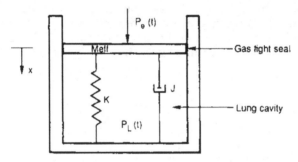

Figure 9.19   *Model of chest cavity (after Ref. 5)*

produces haemorrhage and possible death.

Lung models have been developed for predicting the pressure-time history in the chest cavity. A typical such system[5] is shown in Figure 9.19.

In this model the lung is reduced to a single degree of freedom system in which $P_e(t)$ is the effective pressure acting on the chest wall producing injury. M_{eff} is effective mass of chest wall, K represents tissue elasticity and J is the damping coefficient associated with tissues. $P_L(t)$ is the pressure generated in the lung cavity. It should be noted that the techniques for establishing effective loads and masses will be discussed in Chapter 10. The results of this sort of analysis, when coupled with data gained from experiments on anthropomorphic dummies and also animal experiments, can be extrapolated so that they are applicable to humans.

Information is available either in the form of graphs of incident overpressure plotted against blast wave positive phase duration or as lethality curves. These latter graphs are essentially pressure-impulse curves which are usually presented with suitably scaled axes as discussed in Section 9.5 above. Examples of overpressure versus duration graphs are given in Figures 9.20a, b and c derived from Reference 6 which relate to lung injury for a 70 kg man with the body in different orientations relative to the blast or adjacent to reflecting surfaces.

An alternative representation of this information is shown on the lethality curves of Figure 9.21 for lung damage in man constructed by Baker *et al.*[4] from Figure 9.20.

The axes used here are scaled pressure $\bar{p}_s = \frac{p_s}{p_o}$ where p_s is peak incident pressure and p_o is ambient (usually atmospheric) pressure and scaled impulse $\bar{i}_s$. This is obtained by defining a scaled positive phase duration $\bar{T} = \frac{t_d p_o^{1/2}}{m^{1/3}}$ where t_d is positive phase duration and m is the mass of the subject. The blast wave is taken as triangular pulse which provides a conservative estimate of impulse. Then, the scaled specific impulse is

$$\bar{i}_s = \tfrac{1}{2}\bar{p}_s\bar{T} = \tfrac{1}{2} \cdot \frac{p_s}{p_o} \cdot \frac{t_d p_o^{1/2}}{m^{1/3}}$$

$$= \frac{\tfrac{1}{2}p_s t_d}{p_o^{1/2} m^{1/3}} = \frac{i_s}{p_o^{1/2} m^{1/3}} \tag{9.26}$$

Using these axes, contours represent different levels of survivability with a higher scaled pressure and impulse allowing fewer survivors. Standardisation of the population allows prediction of survival levels for different sectors of humans. Thus, the most vulnerable human is standardised as a 5 kg baby, the next is a 25 kg child followed by a 55 kg adult and finally a 70 kg adult male is the least vulnerable target.

9.7.2 Ear damage

Other directly pressure attributable injuries are to the aural cavities. The mechanism of ear damage can be summarised briefly in the following manner. The ear can respond to noise in the range 20 to 20 000 Hz. However, it cannot respond faithfully to pulses of less than about 0.3 ms duration. If the ear is loaded by a train of such pulses the eardrum will attempt to respond by making a single, large displacement and damage often results. The graph of Figure 9.22 derived from Reference 7 presents overpressure versus duration information relating to eardrum damage.

Such damage can also be presented on pressure-impulse diagrams such as Figure 9.23, developed from a number of sources and presented in Reference 4, though the impulse levels for damage are not so well established as for other structural systems.

Thus, it will be seen that 50% of eardrums will rupture at pressures of about 103 kPa (15 psi) overpressure (= 195 db) while at pressure levels below this temporary hearing loss can occur. Thus, the lowest iso-damage curve (quasi-static asymptote at 160 db) is that for temporary threshold shift (TTS$_1$) which is characterised by temporary deafness.

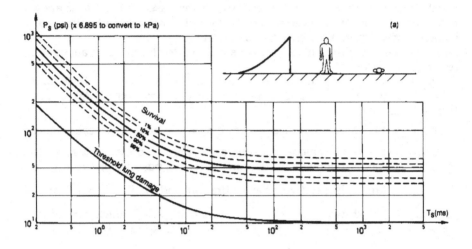

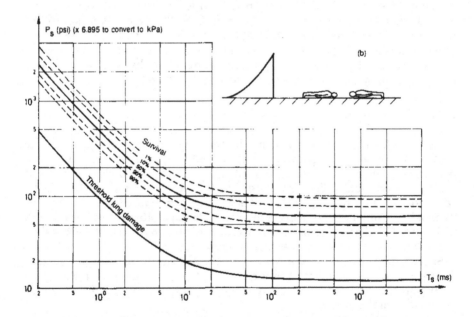

Figure 9.20　(a) *Survival curves for man remote from reflecting surface – long axis of body perpendicular to direction of travel of blast; (b) survival curves for man remote from reflecting surface – long axis of body parallel to direction of travel of blast;*

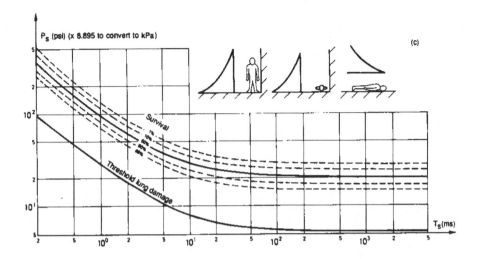

Figure 9.20 (continued) *(c) survival curves for man near reflecting surfaces (after Ref. 6)*

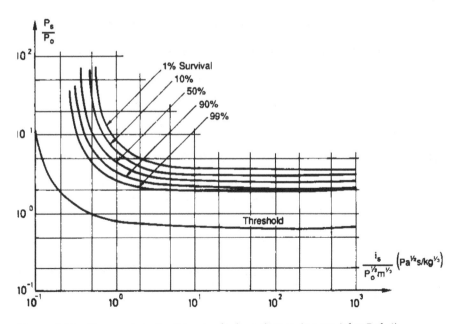

Figure 9.21 *Pressure-impulse diagrams for lung damage in man (after Ref. 4)*

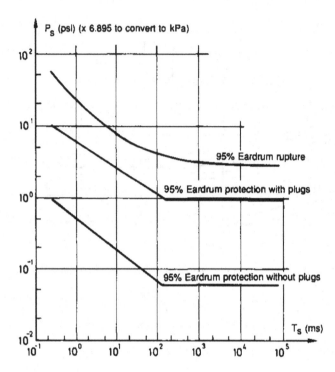

Figure 9.22   *Eardrum rupture curves for man (after Ref. 7)*

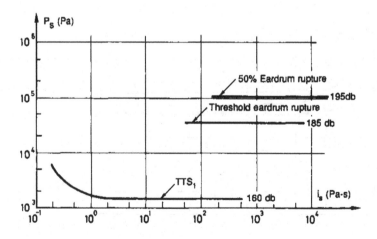

Figure 9.23   *Pressure-impulse diagrams for eardrum rupture (after Ref 4)*

9.7.3 Secondary and tertiary injuries

The discussion above indicates that the body is remarkably resilient to the direct effects of blast overpressure and consequently it could well be that it is indirect effects that are the primary causes of injury to humans.

Secondary injury assessment is not by means of pressure-impulse diagrams, but data are presented here in an alternative form which has a similar, general structure. Therefore, for secondary injuries occasioned by fragments (from, for example, cased weapons) or other debris picked up and carried by the blast wave to impact on the body, iso-casualty curves are available for different debris weights impacting at different velocities as shown in Figures 9.24 and 9.25 extracted from Ahlers[8] and summarised in Reference 9. These figures show 'kill probabilities' for impacts to the head and the body respectively.

Comparison of the two sets of data shows that, for a given impact velocity to produce the same level of injury, a lighter piece of debris is needed in head impact compared to impact on the body. Alternatively, for a given weight of debris to produce the same kill probability, a lower impact velocity is needed if the head is impacted compared with body impact.

In the case of tertiary injuries, these will occur mainly when the body suffers high decelerations, particularly on landing, the worst case being when the head impacts on hard surfaces. The seriousness of the injury can be correlated with the velocity of impact. Pressure-impulse diagrams have been derived from an analysis which models the human body as a cylinder

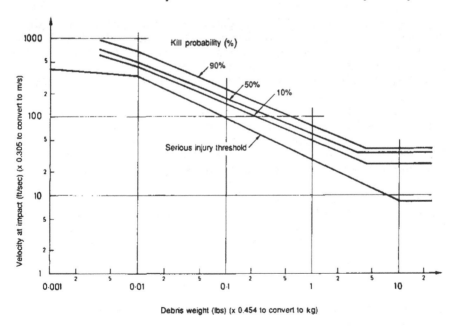

Figure 9.24 *Secondary blast injury: kill probability from fragment impact to the head (after Refs. 8 and 9)*

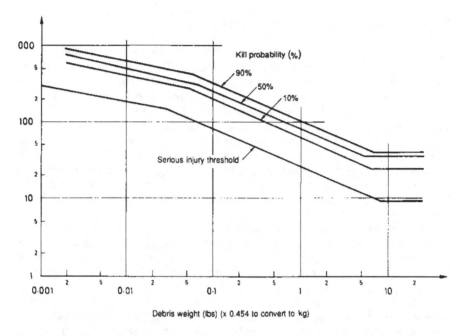

Debris weight (lbs) (x 0.454 to convert to kg)

Figure 9.25 *Secondary blast injury: kill probability from fragment impact to the body and limbs (after Refs. 8 and 9)*

and which is made of material with a density of 1000 kg/m³ with a length to diameter ratio of 5.5, a drag coefficient of 1.3 (summarised in both Refs. 4 and 9). The four body classes as detailed above are used to categorise the population and the effects of different ambient pressure conditions characterised by consideration of standard altitudes of 0, 2000, 4000 and 6000 m above sea level.

Analysis indicates that translation velocity can be represented as a function of incident overpressure p_s and a form of scaled impulse $i_s/m^{1/3}$. A graphical representation of results of analysis is given in Figures 9.26 and 9.27 which are pressure-impulse plots for cases when the subject lands on his skull and for impact of the whole body for sea-level conditions.

On each graph the four lines represent levels of lethality. The lowest represents the 'mostly safe' limit, the next the threshold of damage and the remaining two 50% and 100% lethality respectively. It will be noted that lower velocities of impact will achieve a given lethality level for skull fracture compared to whole body impact.

Example 9.2
The human frame is remarkably resilient and can survive remarkably high overpressure and impulse levels. To illustrate this, consider the specific example of a 70 kg male at sea-level conditions at a range of only 3 m from a 27 kg TNT charge. His scaled distance from the charge is $Z = \frac{R}{W^{1/3}} = \frac{3}{27^{1/3}} = 1$. Using blast parameter vs scaled

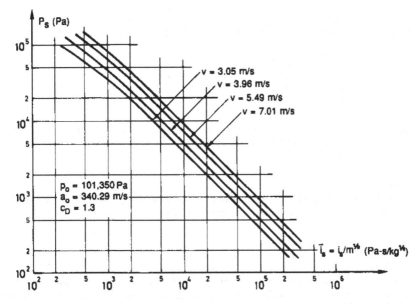

Figure 9.26   *Tertiary blast injury: pressure-impulse diagrams for skull fracture (sea-level conditions) (after Ref. 4)*

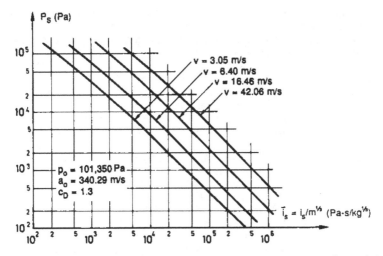

Figure 9.27   *Tertiary blast injury: pressure-impulse diagrams for lethality from whole body translation (sea-level conditions) (after Ref. 4)*

distance graphs (Figure 3.8) we can obtain $p_s = 9 \times 10^5$ Pa and $i_s/W^{1/3} = 160$ so that $i_s = 480$ Pa-s. Now, forming scaled overpressure and impulse we obtain

$$\bar{p}_s = \frac{p_s}{p_0} = \frac{9 \times 10^5}{1.01 \times 10^5} = 8.91$$

$$\bar{i}_s = \frac{480}{[1.01 \times 10^5]^{1/2} \, 70^{1/3}} = 0.366 \tag{9.27}$$

with p_0 as 1.01×10^5 Pa. Now, by reading the iso-casualty curve we find there is an approximately 99% survival chance for lung damage which is a remarkably high figure given the severity of the loading.

Example 9.3

Using the data for Example 9.1 above with $p_s = 900$ kPa and $i_s = 480$ Pa-s, Figure 9.23 indicates a greater than 50% chance of eardrum rupture.

Example 9.4

A 155 mm artillery shell bursts on impact with the ground. From Reference 3, fragments produced by this munition have a mass of typically about 0.065 kg. The initial velocity of these fragments is approximately 1000m/s. When air drag has acted on them their velocity reduces and at a range of about 300m it has fallen to about 60 m/s. What will be the effect on a human target if one of these fragments impacts on the head or on the torso of the subject?

By reference to Figure 9.24 for head impact there is approximately a 90% kill probability. By reference to Figure 9.25 for body impact this kill probability has reduced to about 10%.

9.8   References

[1] Jarrett D.E., Derivation of British Explosives Safety Distances. *Annals of the New York Academy of Sciences*, Vol. 152, Art. 1, pp. 18–35 (1968)

[2] TM5-855-1 Fundamentals of Protective Design for Conventional Weapons. *US Department of the Army Technical Manual* (1985)

[3] CONWEP: Conventional Weapons Effects Program. Prepared by D.W. Hyde, US Army Waterways Experiment Station, Vicksburg (1991)

[4] Baker W.E., Cox P.A., Westine P.S., Kulesz J.J. and Strehlow R.A., *Explosion Hazards and Evaluation*. Elsevier, Amsterdam (1983)

[5] Clemedson C.-J., Jönsson A., Effects of the frequency content in complex air shock wave on lung injury in rabbits. *Aviation, Space and Environmental Medicine* Vol. 47, No. 11, pp. 1143–52 (1976)

[6] Bowen I.G., Fletcher E.R., Richmond D.R., Estimates of man's tolerance to the direct effects of airblast. Technical Progress Report DASA-2113, Defense Atomic Support Agency, US Department of Defense (1968)

[7] Richmond D.R., Yelverton J.T., Fletcher E.R., New airblast criteria for man. Minutes of the 22nd Explosives Safety Seminar, Department of Defense Explosives Safety Board, USA, Anaheim California (1986)

[8] Ahlers E.B., Fragment hazard study. Minutes of the 11th Explosives Safety Seminar Vol. 1, Armed Services Explosives Safety Board, Washington, DC (1969)

[9] Baker W.E., Westine P.S., Kulesz J.J., Wilbeck J.S. and Cox P.A., *A Manual for the Prediction of Blast and Fragment Loading on Structures*. DOE/TIC-11268 US Department of Energy, Amarillo, Texas (1980)

[10] Baker W.E., Kulesz J.J., Richer R.E., Bessey R.L., Westine P.S., Parr V.B. and Oldham G.A., Workbook for Predicting Pressure Wave and Fragment Effects of Exploding Propellant Tanks and Gas Storage Vessels. NASA CR-134906 NASA Lewis Research Centre (1975, reprinted September 1977)

[11] TM5-1300 Design of Structures to Resist the Effects of Accidental Explosions. *US Department of the Army Technical Manual* (1991)

Symbols

F	peak blast load
i_r	peak reflected specific impulse
i_s	peak side-on specific impulse
$\bar{i}_s$	scaled impulse
I	impulse
J	damping coefficient of chest-lung system
K	stiffness
K'	empirical constant
KE	kinetic energy
m	mass of human subject
M	structure mass
M_{eff}	effective mass of chest wall
n	ratio of maximum dynamic displacement to maximum elastic displacement
p	pressure
$P_e(t)$	effective pressure loading chest wall
$P_L(t)$	pressure generated in lung cavity
p_m	maximum overpressure of gas or dust explosion
p_o	atmospheric pressure
p_s	peak side-on overpressure
$\bar{p}_s$	scaled pressure
R	range
R_B	radius for Category B damage
$R(t)$	structure resistance
R_{max}	maximum structure resistance
t	time
t_d	duration of idealised triangular blast load
t_m	time to reach maximum dynamic displacement
t_r	time to reach maximum pressure in gas or dust explosion
T	natural period of vibration of structure
$\bar{T}$	scaled blast wave duration
U	strain energy
W	charge mass
WD	work done by blast load
x	displacement
x_{max}	maximum dynamic displacement

x_{el}	maximum elastic displacement
x_{st}	static displacement
Z	scaled distance
ω	natural circular frequency of vibration

10 Structural response: equivalent systems

10.1 Introduction

In Chapter 9 the principles of using pressure-impulse (P-I) or iso-damage curves as a means of assessing the response of a range of targets were discussed. In this chapter the technique will be extended to consider the behaviour of specific structural elements by considering their response to quasi-static and impulsive loading (which can lead to the development of P-I diagrams). However, it will be seen that having developed such curves an even more convenient representation of response can be derived by converting these distributed mass, flexible systems to more tractable, equivalent, single degree of freedom (SDOF) structures by the evaluation of load and mass factors.

10.2 Energy solutions for specific structural components

10.2.1 Elastic analysis

The approach adopted here is essentially the Rayleigh–Ritz method of analysis[1] and is suitable for specific structural members in an uncoupled analysis. This means that the response of structural elements to a blast load may be considered in isolation assuming that the support conditions are essentially rigid. In the implementation described here the approach produces worst cases of response rather than displacement-time histories.

There are well-defined steps to the solution of a particular structure using this method. Firstly, a mathematical representation of the deformed shape is selected for the structure which satisfies all the necessary boundary conditions relating to displacement. Then, by operating on the deformed shape, the curvature and hence strain of deformation is obtained from which strain energy per unit volume of material can be evaluated. By integration for the whole element the total strain energy can be calculated.

Consideration must now be given to the nature of the blast load impinging on the structure. If the loading is determined as being impulsive then a calculation of total kinetic energy delivered to the structure is made. If, however, the load is quasi-static in nature the work done by the load is found by considering the work done on a small element of the structure then integrating over the loaded area. In the impulsive realm response is evaluated by equating the kinetic energy acquired to the strain energy produced in the structure. In the quasi-static realm response is assessed by equating the work done by the load to strain energy. Having done this it is possible to quantify particular aspects of response such as maximum displacement, maximum strains and maximum stresses.

Example 10.1
To illustrate the technique consider the specific example of a rectangular section slender cantilever of breadth b and depth d that is loaded impulsively by a blast wave delivering a specific reflected overpressure impulse i_r and responds elastically. The structure is shown in Figure 10.1.

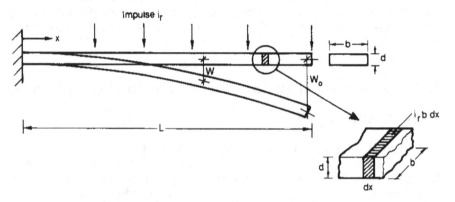

Figure 10.1   *Impulsively loaded elastic cantilever*

We require to calculate the maximum deflection and maximum bending strain experienced by the structure. Firstly, a deformed shape must be selected that satisfies all the boundary conditions. Here the deformed shape under a statically applied uniformly distributed loading is selected though a sine waveform, a parabola or another representation could suffice equally well.

$$W = \frac{W_o}{3}\left[6\left(\frac{x}{L}\right)^2 - 4\left(\frac{x}{L}\right)^3 + \left(\frac{x}{L}\right)^4\right] \tag{10.1}$$

In Equation 10.1, W is the displacement at a distance x from the built-in end of the cantilever of length L. At the free end displacement is W_o. If the cantilever is slender it may reasonably be assumed that the deformed structure acquires strain energy due only to bending. Thus, it is appropriate to use the elastic beam bending equation

$$\frac{M}{I} = \frac{E}{R} = \frac{\sigma}{y} \tag{10.2}$$

where M is bending moment, I is second moment of area of the section about the neutral axis and E is Young's modulus for the beam material. R is the radius of curvature given approximately by the equation

$$R \approx \frac{1}{\left[\dfrac{d^2W}{dx^2}\right]} \tag{10.3}$$

and σ is the direct stress at a distance y from the neutral axis of the section. In this case

$$\frac{1}{R} = \frac{4W_0}{L^2}\left[1 - 2\left(\frac{x}{L}\right) + \left(\frac{x}{L}\right)^2\right] \tag{10.4}$$

Thus we can write

$$M = \frac{EI}{R} = EI\frac{d^2W}{dx^2} = \frac{4EIW_0}{L^2}\left[1 - 2\left(\frac{x}{L}\right) + \left(\frac{x}{L}\right)^2\right] \tag{10.5}$$

The strain energy in bending U can now be evaluated. Firstly, the strain energy per unit volume (dU/dV) can be written for an element of the beam of length dx as

$$\frac{dU}{dV} = \int \sigma d\epsilon = \int \frac{\sigma}{E} d\sigma = \frac{\sigma^2}{2E} = \frac{M^2 y^2}{2I^2 E} \tag{10.6}$$

If the element has cross-sectional area dA and is of length dx then elemental volume $dV = dAdx$ and summing up over the entire cross-section U can be written

$$U = \int \frac{M^2 y^2}{2I^2 E} dAdx = \int \frac{M^2 I}{2I^2 E} dx = \int \frac{M^2}{2EI} dx \tag{10.7}$$

since by definition

$$\int y^2 dA = I \tag{10.8}$$

If the integration is performed over the entire beam length L

$$\begin{aligned}
U &= \int_0^L \frac{M^2}{2EI} dx = \int_0^L \frac{E^2 I^2}{2EI}\left[\frac{d^2W}{dx^2}\right]^2 dx \\
&= \frac{EI}{2} \int_0^L \left[\frac{d^2W}{dx}\right]^2 dx = \frac{8EIW_0^2}{L^4} \int_0^L \left[1 - 2\left(\frac{x}{L}\right) + \left(\frac{x}{L}\right)^2\right]^2 dx \\
&= \frac{8EIW_0^2}{5L^3}
\end{aligned} \tag{10.9}$$

Since it is here assumed that the cantilever is being loaded impulsively, the kinetic energy delivered is evaluated by considering a small element of the beam as shown in

Figure 10.1. By summing up the kinetic energy acquired by each element the total kinetic energy KE is found as

$$KE = \sum_{\text{beam}} \tfrac{1}{2} \times \text{mass of element} \times (\text{initial velocity})^2$$
$$= \int_0^L \tfrac{1}{2}[\rho\, bd\, dx]\left[\frac{i_r b\, dx}{\rho\, bd\, dx}\right]^2 = \int_0^L \tfrac{1}{2}\frac{bi_r^2}{\rho d}\, dx$$

$$= \frac{bi_r^2 L}{2\rho d} \tag{10.10}$$

where the initial velocity is obtained as the ratio of the impulse delivered to each element to the mass of the element. Having gained this energy the structure undergoes deformation converting it into strain energy. Thus, equating kinetic energy and strain energy of bending, we have

$$\frac{bi_r^2 L}{2\rho d} = \frac{8EIW_0^2}{5L^3} \tag{10.11}$$

For a rectangular-section beam, second moment of area about the neutral axis can be written

$$I = \frac{bd^3}{12} \tag{10.12}$$

so that

$$\frac{bi_r^2 L}{2\rho d} = \frac{8Ebd^3 W_0^2}{5 \times 12L^3} \tag{10.13}$$

which can be rearranged to give

$$\frac{W_0}{L} = \frac{\sqrt{15}}{2}\left[\frac{L}{d}\right]\left[\frac{i_r}{d\sqrt{E\rho}}\right] \tag{10.14}$$

This result indicates that, for a given beam, the maximum displacement is proportional to the impulse delivered. From this result the maximum strain (which occurs at the outermost fibres of the cantilever at $y = d/2$) can be evaluated. So, since

$$\frac{M}{I} = \frac{\sigma}{y} = \frac{\epsilon E}{y} \tag{10.15}$$

and with maximum strain $\epsilon_{\max}$ occurring at a distance $d/2$ from the neutral axis then

$$\epsilon_{\max} = \frac{Md}{2EI} = \frac{d^2W}{dx^2}\cdot\frac{d}{2} = \frac{2dW_0}{L^2}\left[1 - 2\left(\frac{x}{L}\right) + \left(\frac{x}{L}\right)^2\right] \tag{10.16}$$

The greatest possible strain will be at the cantilever root. Thus, when $x = 0$, strain is $\epsilon_{x=0}$ given by

$$\epsilon_{x=0} = \frac{2dW_o}{L^2}$$

$$= \frac{2d}{L} \frac{\sqrt{15}}{2} \left[\frac{L}{d}\right] \left[\frac{i_r}{d\sqrt{E\rho}}\right]$$

$$= \sqrt{15} \frac{i_r}{d\sqrt{E\rho}} \tag{10.17}$$

This strain is not a function of length. If the beam is (say) doubled in length then, although double the amount of kinetic energy is delivered, there is double the amount of material able to absorb this energy.

Example 10.2
Consider the same cantilever structure now loaded quasi-statically. A similar analysis can be carried out but now it is assumed that the blast pressure on the structure remains steady during structure deformation: the peak reflected overpressure p_r remains constant. The evaluation of strain energy is as before but now the maximum possible work done by the blast load is calculated by considering the work done WD on a beam element and integrating over the length of the beam

$$WD = \int_0^L p_r b dx W \tag{10.18}$$

substituting the expression for displacement gives WD as

$$WD = \int_0^L \frac{p_r bW_o}{3} \left[6\left(\frac{x}{L}\right)^2 - 4\left(\frac{x}{L}\right)^3 + \left(\frac{x}{L}\right)^4\right] dx \tag{10.19}$$

which on integration gives

$$WD = \frac{2p_r bLW_o}{5} \tag{10.20}$$

Now, work done and strain energy can be equated

$$\frac{8EIW_o^2}{5L^3} = \frac{2p_r bLW_o}{5} \tag{10.21}$$

By rearranging and substituting the expression for second moment of area of a rectangular section beam, maximum displacement, expressed as W_o/L, can be obtained as

$$\frac{W_o}{L} = 3\left[\frac{L}{d}\right]^3 \left[\frac{p_r}{E}\right] \tag{10.22}$$

The maximum strain at the cantilever root $\epsilon_{x=0}$ is given by

$$\epsilon_{x=0} = \frac{2dW_o}{L^2} = \frac{2d}{L} \times 3\left[\frac{L}{d}\right]^3 \left[\frac{p_r}{E}\right]$$

$$= 6\left[\frac{L}{d}\right]^2 \left[\frac{p_r}{E}\right] \tag{10.23}$$

where $\epsilon_{x=0}$ is seen to be a strong function of L. This is because the greater the beam length the greater the 'lever arm' on which the blast load will act to produce strain in the structure.

10.2.2   Influence of deflected shape selection

The exact values of the coefficients in the strain and kinetic energy and work calculations depend on the choice of deflected structure shape. Provided that a shape that satisfies boundary conditions is used, results will be very similar. To illustrate this, the same structure has been analysed for other deflected shapes. The results for a sine curve and a parabola are compared in Table 10.1 below.

Table 10.1

	Static loading	*Sine*	*Parabola*
Coefficient A in $$U = A\left[\frac{EIW_0^2}{L^3}\right]$$	1.6	1.52	2.00
Coefficient B in $$\frac{W_0}{L} = B\left[\frac{L}{d}\right]\left[\frac{i_r}{d\sqrt{E\rho}}\right]$$	1.94	1.99	1.73
Coefficient C in $$\frac{W_0}{L} = C\left[\frac{L}{d}\right]^3\left[\frac{p_r}{E}\right]$$	3.00	2.86	2.00

It would appear that the analyses based on the static deformed shape and the sine curve approximations produce very similar results while the parabolic representation, though easy to apply, may not provide as accurate an assessment, though even in this case differences are not very large.

10.3   Plastic analysis

The approach to the analysis of structural elements that deform beyond their elastic limits proceeds in a similar way to that adopted above for elastic response. To illustrate the approach consider a specific example adapted from Reference 6.

Example 10.3
Consider a simply-supported beam made of material with stress-strain characteristics idealised as 'rigid-plastic' as shown in Figure 10.2.
   The first step is to choose a suitable deflected shape, in this case a parabola in which the deflection W at a distance x measured from the centre of the beam is given by

$$W = W_0\left[1 - \frac{4x^2}{L^2}\right] \tag{10.24}$$

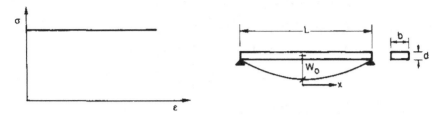

Figure 10.2 *Rigid-plastic simply supported beam*

where the deflection at the centre is W_o. From this we can obtain an expression for the beam curvature R as

$$\frac{1}{R} = \frac{d^2W}{dx^2} = -\frac{8W_o}{L^2} \tag{10.25}$$

From geometrical considerations the rotation $d\theta$ of a beam element of length dx is

$$d\theta = \frac{dx}{R} = \frac{d^2W}{dx^2}dx \tag{10.26}$$

The strain energy of bending U is found by summing up the work done by the bending moment on each element of the beam (here always M_y) as it causes the elemental rotation $d\theta$

$$U = 2\int_0^{L/2} M_y d\theta$$

$$= 2\int_0^{L/2} M_y \frac{d^2W}{dx^2}dx$$

$$= \frac{16M_yW_o}{L^2}\int_0^{L/2}dx = \frac{8M_yW_o}{L} \tag{10.27}$$

where M_y is the moment to cause yielding. Consider for this example that the beam is loaded impulsively. Assuming a rectangular section the kinetic energy delivered will be as before in Example 10.1. Thus, equating kinetic energy and strain energy

$$\frac{8M_yW_o}{L} = \frac{i_r^2bL}{2\rho d} \tag{10.28}$$

For a rectangular section beam in a plastic state, from Figure 10.3 it can be shown that the yield moment is

$$M_y = \sigma_y\frac{bd}{2} \times \frac{d}{2} = \frac{\sigma_y bd^2}{4} \tag{10.29}$$

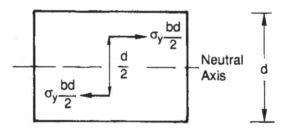

Figure 10.3   *Evaluation of yield moment in plastic section*

Thus we can write

$$\frac{8\sigma_y bd^2 W_0}{4L} = \frac{i_r^2 bL}{2\rho d}$$

$$\therefore \quad \frac{W_0}{L} = \frac{i_r^2 L}{4\sigma_y \rho d^3}$$

(10.30)

If instead of being simply-supported the beam was clamped at each end, similar analysis yields

$$\frac{W_0}{L} = \frac{i_r^2 L}{8\sigma_y \rho d^3}$$

(10.31)

Indeed, any support conditions can be accounted for, the general result being given by

$$\frac{W_0}{L} = \frac{i_r^2 L}{N\sigma_y \rho d^3}$$

(10.32)

where N is between 4 and 8.

As noted above, the choice of deformed shape is of relatively minor importance in the analysis provided boundary conditions are satisfied. Of more significance is the coupling between the structural component and its support. The approach outlined here will be satisfactory though, provided the supporting structure period is a lot greater than the vibration period of the component being analysed. If the support must be taken as exhibiting flexibility in a timescale commensurate with the component then the analysis is necessarily more complex. The method above generates information that can, of course, be represented in a pressure-impulse diagram and is equally applicable to plates, columns, strips and other components as the examples of Figures 10.4 and 10.5 indicate.

In Figure 10.5 the suffixes to the plate shape factors ϕ indicate whether the loading is impulsive (*i*), quasi-static or pressure (*p*) and whether the plate material is brittle (*b*) meaning that failure occurs when stress reaches yield at one location only or ductile (*d*) meaning that failure occurs only after a complete failure mechanism (the appropriate yield line pattern) has been generated.

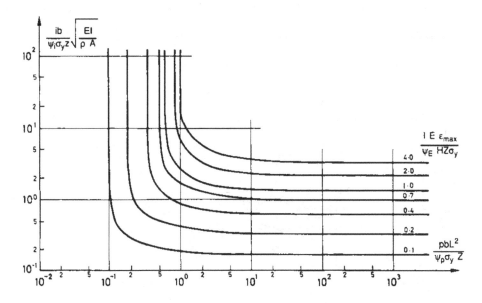

Boundary conditions	ψ_p	ψ_i	ψ_E	$\psi_{w_o}\left(=\dfrac{H_{w_o}}{L^2\,\varepsilon_{max}}\right)$
S S	10.000	0.913	1.250	0.2083
C C	23.100	0.861	1.925	0.0625
C P	15.830	0.885	1.979	0.0867
Cantilever	3.333	0.577	1.000	0.5000

S : simple support C : clamped P : pinned

Figure 10.4 *Pressure-impulse diagrams for elastic-plastic beams (after Ref. 4)*

10.4 Lumped-mass equivalent single degree of freedom systems

The complete analyses presented above are valuable though, for more complex structural elements and load configurations, implementation of this approach could be rather time-consuming. To aid assessment of response, the behaviour of complex structures can be approached by representing the structure as an SDOF lumped-mass system – a so-called equivalent system. The equation of motion so derived will be of the general form introduced in Chapter 8 and developed in Chapter 9 and can be solved either analytically or (more usually) numerically (as discussed in Chapter 11) to obtain a

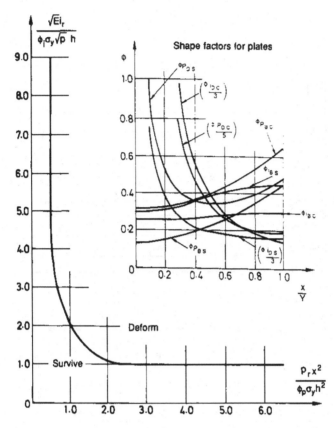

Figure 10.5 *Pressure-impulse diagrams for plates (after Ref. 4)*

deformation-time history for the structure. Often, though, as described above, maximum displacement may be all that is required. This approach, though failing to provide detailed aspects of response, allows a good insight into important features of behaviour and will, furthermore, give an over-assessment of response. This conservative analysis, therefore, has attraction for the designer.

10.4.1 Equation of motion for an SDOF system

The equation obtained in Chapter 8 to describe an SDOF structure was

$$M\ddot{x} + c\dot{x} + Kx = P(t) \tag{10.33}$$

This equation is usually simplified by setting the damping term c to zero because in one cycle of displacement attenuation is small, ignoring c is a conservative approach and for structures taken beyond their elastic limit energy dissipation is mainly by plastic deformation. The damping coefficient is anyway usually unknown though it is sometimes acceptable to include a damping factor of up to 10% which can have benefits in material costs. Thus, for the system above with c set to zero and writing the spring resistance term as a more general resistance function $R(x)$ gives

$$M\ddot{x} + R(x) = P(t) \tag{10.34}$$

In creating an equivalent SDOF structure it must be realised that real structures are multi-degree of freedom systems where every mass particle has its own equation of motion. Thus, to simplify the situation it is necessary to make assumptions about response and in particular characterise deformation in terms of a single point displacement. Examples are shown in Figures 10.6a and b where in the SDOF system the suffix e means equivalent.

The method relies, as in the examples above, on considering the energies of the real structure and the equivalent system and equating them. This means that, by ensuring equal displacements and velocities in the two systems, kinematic similarity is maintained. The complete energy relationship may be written as

$$WD = U + KE \tag{10.35}$$

In the derivations of impulsive and quasi-static response described above it was this equation that was effectively reduced to two terms for the two extremes of behaviour. When assessing quasi-static response, work done by

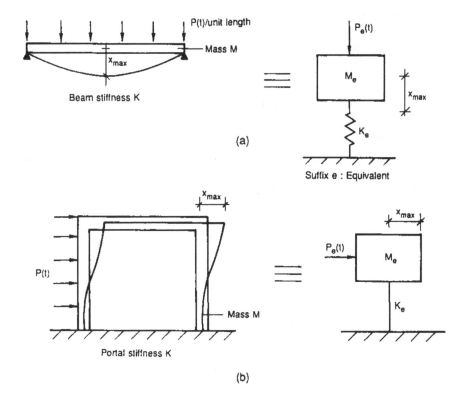

Figure 10.6 *Examples of equivalent systems; (a) beam or one-way slab; (b) portal frame*

the load (WD) was equated to internal strain energy (U) and the kinetic energy term (KE) was taken as negligibly small. For impulsive loading WD is set to zero and kinetic energy delivered is converted to structure strain energy.

Example 10.4.
As an example of the approach, consider a simply-supported beam responding elastically as shown in Figure 10.6a above. Firstly, a suitable deformed shape is assumed: here the displacement W at distance x from the end of the beam is given by

$$W = \frac{16}{5L^4} (L^3 x - 2Lx^3 + x^4) W_0 \qquad (10.36)$$

The evaluation of work done, strain energy and kinetic energy for the beam is given below.

$$WD = \int_0^L p(t) W(x)\, dx = \int_0^L \frac{16p(t)}{5L^4} (L^3 x - 2Lx^3 + x^4) W_0\, dx$$

$$= \frac{16}{25} p(t) L W_0 = \frac{16}{25} P(t) W_0 \qquad (10.37)$$

$$U = \int_0^L \frac{M^2}{2EI}\, dx = \frac{EI}{2} \int_0^L \left[\frac{d^2 W}{dx^2} \right]^2 dx$$

$$= \frac{EI}{2} \int_0^L \left[\frac{192 W_0}{5L^2} \right]^2 (x^2 - Lx)^2 dx = \frac{24.58 EI W_0^2}{L^3} \qquad (10.38)$$

$$KE = \frac{1}{2} \int_0^L \rho A [\dot{W}(x)]^2 dx = \frac{1}{2} \rho A \int_0^L \frac{256 \dot{W}_0^2}{25L^8} [L^3 x - 2Lx^3 + x^4]^2 dx$$

$$= 0.252 \rho A L \dot{W}_0^2 \qquad (10.39)$$

where $p(t)$ is the distributed blast load and $P(t)$ is the total blast load. The equivalent system will be as shown in Figure 10.6a which will have the same maximum displacement W_0 and maximum initial velocity $\dot{W}_0$ as for the real structure. The evaluation of WD, U and KE is much simpler in this case

$$W = P_e(t) W_0$$

$$U = \frac{1}{2} K_e W_0^2$$

$$KE = \frac{1}{2} M_e \dot{W}_0^2 \qquad (10.40)$$

Equating these quantities to those of the real system gives for the work terms

$$\frac{16}{25} W_0 P(t) = P_e(t) W_0 \qquad (10.41)$$

We can thus define a load factor K_L as

$$K_L = \frac{P_e(t)}{P(t)} = \frac{16}{25} = 0.64 \qquad (10.42)$$

Equating strain energy for the two systems gives

$$\frac{24.58EIW_o^2}{L^3} = \frac{1}{2}K_e W_o^2 \tag{10.43}$$

For a simply-supported elastic beam the central deflection W_o, under a point load P is given by

$$W_o = \frac{5PL^3}{384EI} \tag{10.44}$$

so a measure of stiffness, the slope of the load-deflection graph is

$$\frac{P}{W_o} = \frac{384EI}{5L^3} = K \tag{10.45}$$

Therefore, substitution of Equation 10.45 into Equation 10.43 and rearrangement allows definition of a stiffness factor K_s given by

$$K_s = \frac{K_e}{K} = \frac{2 \times 5 \times 24.58}{384} = 0.6401 \tag{10.46}$$

Equating kinetic energy for the two systems gives

$$0.252\rho AL\dot{W}_o^2 = \frac{1}{2}M_e\dot{W}_o^2 \tag{10.47}$$

Since the term ρAL is the mass M of the beam, a mass factor K_M can be defined as

$$K_M = \frac{M_e}{M} = 0.252 \times 2 = 0.504 \tag{10.48}$$

If these calculations had been based on a sine curve as the deformed shape almost identical answers would have been obtained. Thus if displacement W at a distance x along the beam is given by

$$W = W_o \sin\frac{\pi x}{L} \tag{10.49}$$

the total work done by the load is

$$WD = \frac{2}{\pi}W_oP(t) \tag{10.50}$$

the strain energy acquired by the structure is

$$U = \frac{\pi^4}{4}\frac{EIW_o^2}{L^3} \tag{10.51}$$

and the kinetic energy gained is

$$KE = \frac{1}{4}\rho AL\dot{W}_o^2 \tag{10.52}$$

If these results are equated to the response of the equivalent system as detailed by Equation 10.40 the load factor K_L is found as 0.6366, the stiffness factor K_s is 0.6345 and the mass factor K_M is 0.5 which should be compared with the values 0.64, 0.504 and 0.6401 obtained previously.

It is worth noting that the load and stiffness factors are very similar and it is general practice to set them equal in the analysis of a structural element. Thus we have

$$K_s = K_L \tag{10.53}$$

10.4.2 Dynamic reactions

One further important calculation that should be made is of the dynamic support reactions generated by the blast load. Again the analysis is best illustrated by taking the specific example of the simply supported beam of Example 10.4.

Example 10.5
For a simple supported beam, take the static deflected shape analysis to illustrate the approach. The deformed shape with appropriate notation is shown in Figure 10.7.

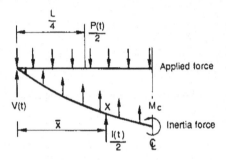

Figure 10.7 *Evaluation of dynamic reactions for elastic beam (after Ref. 5)*

For the half beam shown, moments are taken about the centre of inertial resistance of the beam as it deforms. This is taken to coincide with the centroid of the shape swept out by the beam as it deforms. For the shape chosen here the centroid can be shown to be at $\bar{x} = 61L/192$ from the end of the beam. Taking moments about this point X gives

$$V(t)\bar{x} - M_c - \frac{1}{2}P(t)\left(\bar{x} - \frac{L}{4}\right) = 0 \tag{10.54}$$

where M_c is the moment at the centre of the beam. If a static load R is needed to cause the same deflection as the blast load then it can be shown that

$$M_c = \frac{RL}{8} \tag{10.55}$$

and hence Equation 10.54 can be rewritten to give the dynamic reaction $V(t)$ as

$$V(t) = 0.393R + 0.107P(t) \tag{10.56}$$

The same analysis can be used to develop load and mass factors for plastically deforming structures as illustrated below.

Example 10.6

A simply supported beam which has undergone yield at the centre is shown in Figure 10.8.

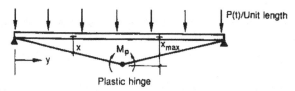

Figure 10.8 *Plastic deformation of simply supported beam*

At a distance x from support the deflection is given by

$$W = \frac{W_0}{L/2} \cdot x \qquad (10.57)$$

The total work done by the load is evaluated as

$$WD = 2 \int_0^{L/2} p(t) \, dx W$$

$$= 2 \int_0^{L/2} p(t) \frac{W_0}{L/2} \cdot x \, dx$$

$$= \frac{2p(t)W_0}{L/2} \left[\frac{x^2}{2} \right]_0^{L/2} = \frac{4p(t)W_0}{L} \cdot \frac{L^2}{8}$$

$$= \frac{p(t)LW_0}{2} = \frac{P(t)W_0}{2} \qquad (10.58)$$

which can be equated to the equivalent system work $(= P_e(t)W_0)$ to give the load factor K_L as

$$K_L = \frac{P_e(t)}{P(t)} = 0.5 \qquad (10.59)$$

Kinetic energy is evaluated as

$$KE = 2 \int_0^{L/2} \frac{1}{2} (m \, dx) \dot{W}_0^2$$

$$= 2 \int_0^{L/2} \frac{1}{2} m \, dx \left[\frac{x}{L/2} \dot{W}_0 \right]^2$$

$$= \frac{4m\dot{W}_0^2}{L^2} \int_0^{L/2} x^2 \, dx$$

$$= \frac{4m\dot{W}_0^2 L^3}{24L^2} = \frac{1}{6} M\dot{W}_0^2 \qquad (10.60)$$

Equating this to the kinetic energy of equivalent system ($= \frac{1}{2} M_e \dot{W}_0^2$) the mass factor K_M is

$$K_M = \frac{M_e}{M} = \frac{1}{3} \tag{10.61}$$

The fully plastic moment of resistance M_p opposes the bending moment due to applied load which, at the beam centre, is

$$M_p = \frac{p(t)L^2}{8} = \frac{P(t)L}{8} \tag{10.62}$$

$$\therefore \quad P(t) = \frac{8M_p}{L}$$

Thus $P(t)$ is constant and, since the equivalent load is $\frac{1}{2}P$ the equivalent resistance K_e is $4M_p/L$.

If this beam is a component of a larger structure, load is transferred to other parts as dynamic reactions. For the present example the situation may be represented as in Figure 10.9.

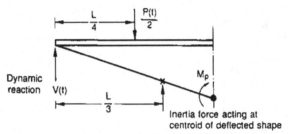

Figure 10.9   *Evaluation of dynamic reactions for plastically deforming beam*

Dynamic reactions $V(t)$ are calculated as for the elastic beam of Example 10.5 above by taking moments about the point where inertial resistance is concentrated at the centroid of the deflected shape of the structure which here will be at $L/3$ from the support since we assume that the beam takes up a linear deflected shape. Therefore, taking moments about point X in Figure 10.9 gives

$$V(t)\frac{L}{3} - M_p - \frac{1}{2}P(t)\left[\frac{L}{3} - \frac{L}{4}\right] = 0 \tag{10.63}$$

where M_p is $R_pL/8$ and where R_p is the magnitude of the static load necessary to cause the hinge. Substitution and rearrangement gives dynamic reactions as

$$V(t) = \frac{3}{L}\left[\frac{R_pL}{8} + \frac{1}{2}P(t)\frac{L}{12}\right] \tag{10.64}$$

$$= 0.375R_p + 0.125P(t)$$

Note that in the static case the reaction at support would be $\frac{1}{2}R_p$.

Table 10.2 Transformation factors for beams and one-way spanning slabs

Loading: Total = F	Strain range	Load factor K_L	Mass factor K_M C	U	Load-mass factor K_{LM} C	U	Maximum resistance $R = aM_P/L$ a	Spring constant $K = bEI/L^3$ b	Dynamic reaction $V = cR + dF$ c	d
SS: uniform	E	0.64		0.50		0.78	8	384/5	0.39	0.11
	P	0.50		0.33		0.66	8	0	0.38	0.12
SS: point (centre)	E	1.00	1.00	0.49	1.00	0.49	4	48	0.78	-0.28
	P	1.00	1.00	0.33	1.00	0.33	4	0	0.75	-0.25
SS: point (third points)	E	0.87	0.76	0.52	0.87	0.60	6	56.4	0.62	-0.12
	P	1.00	1.00	0.56	1.00	0.56	6	0	0.52	-0.020
FF: uniform	E	0.53		0.41		0.77	12	384	0.36	0.14
	E/P	0.64		0.50		0.78	8*	384/5	0.39	0.11
	P	0.50		0.33		0.66	8*	0	0.38	0.12
FF: point	E	1.0	1.0	0.37	1.0	0.37	4*	192	0.71	-0.21
	P	1.0	1.0	0.33	1.0	0.33	4*	0	0.75	-0.25
SF: uniform	E	0.58		0.45		0.78	8	185	0.26	0.12#
									0.43	0.19##
	E/P	0.64		0.50		0.78	4**	384/5	0.39	0.11-
	P	0.50		0.33		0.66	4**	0	0.38	0.12-
SF: point	E	1.0	1.0	0.43	1.0	0.43	16/3	107	0.54	0.14#
									0.25	0.07##
	E/P	1.0	1.0	0.49	1.0	0.49	2**	48	0.78	-0.28-
	P	1.0	1.0	0.33	1.0	0.33	2**	0	0.75	-0.25-
SF: point (third points)	E	0.81	0.67	0.45	0.83	0.55	6	132	0.17	0.17#
									0.33	0.33##
	E/P	0.87	0.76	0.52	0.87	0.60	2***	56	0.62	-0.12-
	P	1.0	1.0	0.56	1.0	0.56	2***		0.52	-0.02-

Key: S Simple support
F Fixed support (encastre)
E Elastic
P Plastic
C Structure considered as concentrated mass with equal parts lumped at load points
U Structure of uniform mass
* M_P replaced by sum of ultimate moment capacity at support and at midspan
** M_P replaced by sum of ultimate moment capacity at support and twice midspan capacity
*** M_P replaced by sum of ultimate moment capacity at support and three times midspan capacity
\# Reaction at simple support
\#\# Reaction at fixed support
\- Reaction given by $cR+dF +/-$ (ultimate moment capacity at support/span)

Table 10.3 Transformation factors for two-way spanning slabs with four sides simply supported

Strain range	a/b	Load factor K_L	Mass factor K_M	Load-mass factor K_{LM}	Maximum resistance $R = \frac{c}{a}(dM_{pfa} + eM_{pfb})$			Spring constant $K = fEI_a/a^2$	$V_A = gF + hR$		$V_B = iF + jR$	
					c	d	e	f	g	h	i	j
E	1.0	0.46	0.31	0.68	12	1	1	252	0.07	0.18	0.07	0.18
	0.9	0.47	0.33	0.70	1	12	11.0	230	0.06	0.16	0.08	0.20
	0.8	0.49	0.35	0.71	1	12	10.3	212	0.06	0.14	0.08	0.22
	0.7	0.51	0.37	0.73	1	12	9.8	201	0.05	0.13	0.08	0.24
	0.6	0.53	0.39	0.74	1	12	9.3	197	0.04	0.11	0.09	0.26
	0.5	0.55	0.41	0.75	1	12	9.0	201	0.04	0.09	0.09	0.28
P	1.0	0.33	0.17	0.51	12	1	1	0	0.09	0.16	0.09	0.16
	0.9	0.35	0.18	0.51	1	12	11.0	0	0.08	0.15	0.09	0.18
	0.8	0.37	0.20	0.54	1	12	10.3	0	0.07	0.13	0.10	0.20
	0.7	0.38	0.22	0.58	1	12	9.8	0	0.06	0.12	0.10	0.22
	0.6	0.40	0.23	0.58	1	12	9.3	0	0.05	0.10	0.10	0.25
	0.5	0.42	0.25	0.59	1	12	9.0	0	0.04	0.08	0.11	0.27

Key:
E   Elastic
P   Plastic
a   Slab width (short side)
b   Slab length (long side)
I_a Moment of inertia per unit width
F   Total load on slab
M_{pfa} Total positive ultimate moment capacity along midspan section parallel to short edge
M_{pfb} Total positive ultimate moment capacity along midspan section parallel to long edge
V_A Total dynamic reaction along short side
V_B Total dynamic reaction along long side

10.5   Load-mass factor

Load and mass factors have been developed for many elements and it is often convenient to define a load-mass factor K_{LM} where

$$K_{LM} = \frac{K_M}{K_L} \tag{10.65}$$

The value of this factor is easily seen. The equivalent system equation of motion can be expressed in terms of mass and load factor term. Division through by load factor (assuming that stiffness and load factors are equal) yields the required result. Thus, for an elastic system

$$M_e \ddot{x} + K_e = P_e(t)$$

$$K_M M \ddot{x} + K_L K x = K_L P(t)$$

$$K_{LM} M \ddot{x} + K x = P(t) \tag{10.66}$$

Tables 10.2 and 10.3 give factors for a variety of structural elements for both elastic and plastic response.

10.6   References

[1] Todd J.D. *Structural Theory and Analysis*. Macmillan, London (1974)

Symbols

A	cross-sectional area of beam
b	breadth of beam
c	damping coefficient
d	depth of beam
E	Young's modulus of elasticity
i_r	specific reflected impulse
I	second moment of area of beam
K	stiffness
K_e	equivalent stiffness
K_L	load factor
K_M	mass factor
K_{LM}	load-mass factor
K_s	stiffness factor
KE	kinetic energy
L	length of beam
m	mass per unit length of beam
M	bending moment, mass of structure
M_c	bending moment at centre of beam
M_e	equivalent mass
M_y	bending moment to cause yielding
M_p	fully plastic bending moment

N	factor relating to end-fixity of beam
p_r	peak reflected overpressure
$P(t)$	blast load
$P_e(t)$	equivalent blast load
P	total load on beam
R	static load to cause same deflection as blast load, radius of curvature
$R(x)$	structure resistance
R_p	static load to cause plastic hinge in beam
U	strain energy of bending
V	structure volume
$V(t)$	dynamic reaction
W	deflection of beam
W_o	maximum deflection of beam
WD	work done by blast load
x	distance along beam
y	distance from beam neutral axis
ϵ	strain
ϵ_{max}	maximum strain in beam
$\epsilon_{x=0}$	strain at $x = 0$ on beam
θ	rotation of beam element
ρ	density of beam material
σ	stress
σ_y	yield stress of material

11 Structural response: incremental solution of equation of motion

11.1 Introduction

In Chapter 10 we developed the concept of a structural element suffering transient deflection under the influence of an imposed blast loading. For the purposes of analysis, the element can be represented as a single degree of freedom system subjected to a transient load which has the value $F(t)$ at a time t. The inherent stiffness of the element provides resistance to deflection. The magnitude of this resistance is denoted by $R(x)$ at deflection x. Applying Newton's second law to the element gives the equation

$$F(t) - R(x) = M\ddot{x} \tag{11.1}$$

In order to predict the transient response of a structural element an incremental solution of Equation 11.1 is developed. Initially, when $x = 0$, the system offers no resistance and so the acceleration of the element can be found from a knowledge of the initial force $F(0)$ and the mass of the element M. Assuming the acceleration to be constant during the first small time increment, the displacement is found and the resistance evaluated at the end of the first increment. The substitution of these values into Equation 11.1, together with the updated value of F, allows the new acceleration to be determined and the process repeated. This incremental approach provides a deflection-time history for the transient displacement of the structural element. The maximum deflection of the element is compared with that which can be tolerated within specified damage criteria to determine whether the element will survive.

This represents the simplest of incremental methods which can be adopted. This chapter describes the incremental approach in more detail and provides advice on the determination of the resistance-deflection function $R(x)$ for specific structural forms.

11.2 Numerical solution of equations of motion.

In the structures that have been analysed in Chapter 10, each has been reduced to a single degree of freedom system and the impulsive and quasi-static bounds of the response have been evaluated. More generally, a structure may be more accurately represented by a number of degrees of freedom. For instance the response of the simple target illustrated in Figure 11.1 can be described by three degrees of freedom, two translatory in the x and z directions and a third rotational degree of freedom θ.

For every degree of freedom an equation of motion can be derived which can be solved to yield components of acceleration, velocity and displacement. Take the specific example of a vehicle idealised as a rigid body (which can have up to six degrees of freedom – three translatory and three rotational) as shown in Figure 11.2.

It is subjected to a force which has both horizontal and vertical components which vary with time. Under the action of these forces the vehicle can move horizontally, vertically and rotate (about centre 0). The system therefore has three degrees of freedom denoted by coordinates x, z and θ and equations of motion for each displacement coordinate must be derived.

$$x : M\frac{d^2x}{dt^2} + \mu Mg = F_x(t)$$

$$z : M\frac{d^2z}{dt^2} + Mg = F_z(t)$$

$$\theta : I\frac{d^2\theta}{dt^2} + MgA = F_x(t)B + F_z(t)C \tag{11.2}$$

where M is the vehicle mass, μ is the coefficient of friction between the vehicle and the ground, g is gravitational acceleration and I is the second

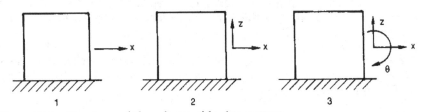

Figure 11.1   *One, two and three degree of freedom systems*

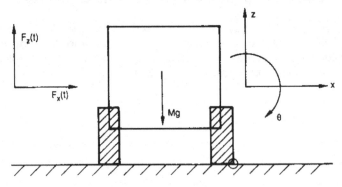

Figure 11.2   *Vehicle target with three degrees of freedom*

moment of area of the body about 0 and A, B and C are the moment arms of the forces acting on the body measured from 0.

11.3   Resistance term

Section 11.4 deals in some detail with resistance functions for specific structural forms. The discussion here is limited to some general observations about the nature of the resistance term in the equation of motion for a particular degree of freedom.

Consider a single degree of freedom system that, when subject to a horizontal force, can only move horizontally as shown in Figure 11.3.

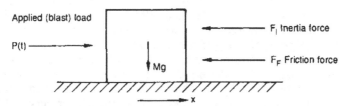

Figure 11.3   *Forces on single degree of freedom system*

By considering the forces acting on the body at a given time the equation of motion states, of course, that the opposing force generated by the structure must equal the applied force

$$F_I + F_F = F(t) \tag{11.3}$$

where F_I is inertial resistance and F_F is a resistance force which could be friction. If the body has mass M and the coefficient of friction between the body and the ground is μ then the equation can be expanded to

$$M\frac{d^2x}{dt^2} + \mu Mg = F(t) \tag{11.4}$$

To prevent excessive horizontal motion the target might be restrained by some sort of cable which tethers the target to a fixed point and the equilibrium equation becomes

$$M\frac{d^2x}{dt^2} + \mu Mg + F_C = F(t) \tag{11.5}$$

If the cable is elastic, the opposing force F_C is proportional to the cable extension and equal to Kx where K is the spring constant. Generally the resistance offered by any restraint depends on the material and on its dimensions. For example, a cable made of mild steel whose load-extension characteristics are as shown in Figure 11.4 displays three different resistances.

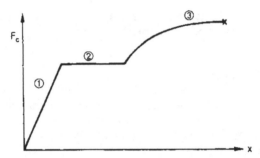

Figure 11.4 *Load-extension characteristics of mild steel cable*

In region 1, behaviour is elastic as described above, while in region 2 the material has yielded and resistance is a different function of extension $f_1(X)$ (say) and in region 3 where the material undergoes work hardening, this relationship changes again and resistance is $f_2(X)$.

The response of a restraint can also depend on how fast the loading is applied (in the case of blast loading this will be very fast). For mild steel, the level of stress (given here by F_C divided by the cross-sectional area of the cable) at which yield occurs can increase by up to 60% for very fast loading rates as shown schematically in Figure 11.5.

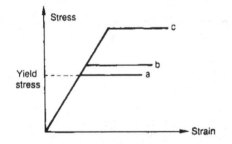

	Time to reach yield stress	Value
a	Several seconds	σy
b	0.1 s	$1.1\,\sigma y$
c	0.01 s	$1.6\,\sigma y$

Figure 11.5 *Strain-rate: effect on yield strength*

In the design of structures to resist blast loading, definition of the structure or structural element resistance function is clearly vital. In some cases the actual resistance function may be too complex to operate with efficiency over the entire displacement range of a particular element and simplification is often undertaken. The example shown in Figure 11.6, taken from the US Army Manual TM5-1300,[1] shows how a resistance function for a concrete structural element exhibiting three steps of elastic response before achieving its ultimate resistance r_u can be approximated for design purposes by an equivalent resistance function in which the same amount of strain energy is absorbed as in the real resistance function.

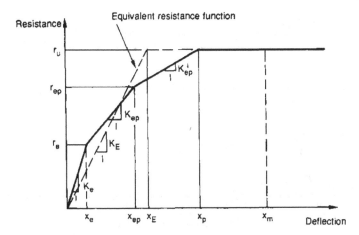

Figure 11.6 *Tri-linear, elastic-plastic, resistance-deflection function*

11.4 Resistance function for specific structural forms

In developing the resistance function for a structural element, it is assumed that the element will offer essentially the same resistance to deflection when deformed dynamically as it will offer when deformed quasi-statically. The only adjustment which is normally incorporated is an enhancement in the ultimate resistance of the element, due to the improvement in strength observed in dynamic loading. In the following sections the general form of the resistance function for some common structural forms will be presented. Detailed design charts are not included but these can be found in TM5-1300[1].

11.4.1 Resistance function – general description

The resistance-deflection function for a structural element is strictly a graph of the uniform pressure which would be necessary to cause deflection at the central point of the element during its transient displacement. In practice a simplified function is used which is a prediction of the resistance which the element would offer in a quasi-static test. Figure 11.7 depicts the resistance function which would be adopted for a reinforced concrete wall with built-in supports along two edges but no support along the other two, i.e. a one-way-spanning slab. The initial part of the graph OA represents the elastic deformation of the slab. At the point A, yield lines develop along the built-in supports, allowing rotation. During this phase, AB, deformations in the central part of the slab are elastic while rotation occurs in the yield lines. The increments of displacement during this *elasto-plastic* phase are estimated from the elastic stiffness of a simply supported one-way-spanning slab. More complicated structural forms, e.g. a two-way-spanning slab with built-in supports, may have two or more elasto-plastic phases as the yield-line system develops progressively in the slab. The phase BC represents the

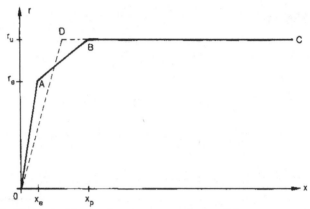

Figure 11.7 *Idealised resistance for a one-way-spanning slab*

ultimate resistance of the element once a sufficient system of yield-lines has been established to form a collapse mechanism in the slab. This *fully plastic* phase will last until failure at one of the supports occurs. For the case of two-way-spanning slabs it is possible for failure to occur along two opposite supports while the other pair of supports remains intact – in which case the slab will continue to offer resistance, but at a lower level. Experience shows that a negligible error is introduced by replacing the resistance function OABC with the bilinear function ODC, provided the area under both functions is the same.

11.4.2   Quantifying the resistance function

In this section, the method by which the coordinates of points A and B in Figure 11.7 are found will be presented. The extension of this method for two-way-spanning slabs will also be described.

Point A, the end of the elastic phase for the one-way-spanning slab, is reached when the bending moment at the supports reaches the ultimate negative moment capacity at the supports. If the distance between supports is L and the slab is subjected to a uniform pressure p, the bending moment per metre run at the supports (Figure 11.8) is given by $p\frac{L^2}{12}$. Thus the pressure to cause yield r_e is $r_e = \frac{12M_N}{L^2}$ where M_N is the ultimate negative moment capacity per metre run of the slab at the supports. The deflection at this stage x_e is found from simple structural analysis to be $\frac{r_e L^4}{384EI}$ where I is the second moment of area per metre run of the slab, assuming tension cracking has taken place, and E is the Young's modulus of concrete.

The steps described above can readily be extended to two-way-spanning slabs. Solutions have been developed, using finite-element analysis, for the elastic bending of slabs of all aspect ratios and support conditions. These are presented in graphical form in Reference 1. Having established the coordinates of A, we now determine r_u and x_p, the coordinates of B. For this simple case r_u is obtained from an upper-bound-plasticity solution (see Section

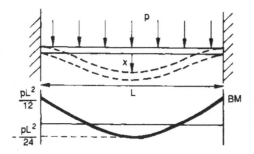

Figure 11.8 *Uniformly loaded, one-way-spanning slab*

13.5.2) in which the work done by the applied pressure is equated to the energy dissipated in the yield lines. This gives

$$r_u = \frac{8}{L^2}(M_N + M_P)$$

where M_N and M_P are, respectively, the ultimate moment capacity per metre run of the slab in negative and positive bending. The increment in deflection during this elasto-plastic phase is obtained by assuming the slab to be simply-supported:

$$x_p - x_e = (r_u - r_e)\frac{5L^4}{384EI}$$

For the case of two-way-spanning slabs, charts are presented in Reference 1 which enable the designer to identify the appropriate collapse mechanism and thereby deduce the ultimate resistance of the slab. Although the resistance function obtained in this way may have more stages, the process of developing the function is intrinsically no more complex than that described above. Recent work by Keating et al.[2] provides resistance functions for wall slabs containing door and window openings.

11.4.3 Resistance function for unreinforced masonry walls

Unreinforced masonry walls have poor resistance to blast loading. However, in assessing existing structures, it may be necessary to examine the response of a masonry building for which a resistance function is required. The resistance offered by a masonry wall comprises two components:

1 The resistance due to elastic slab action up until the point when the mortar on the tension face cracks.
2 The restoring moment due to the self-weight of the slabs.

Figure 11.9 shows the cross-section of a masonry wall supported against horizontal movement along top and bottom edges. No restraint is provided against vertical movement along the upper edge. Fracture has occurred along a horizontal joint at a height h_2 above the ground. The inner edge of the fracture surface has translated a distance y and the system experiences a restoring moment up until the point where y equals w, the thickness of the

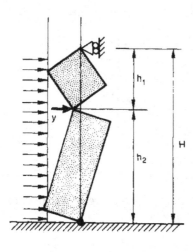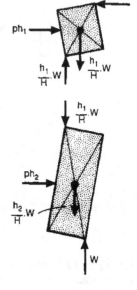

Figure 11.9   *Collapse mechanism for a masonry wall*

wall. By taking moments for the upper segment above the top edge we obtain

$$p = \frac{W(w - y)}{Hh_1} \tag{11.6}$$

where W is the weight per metre run of wall.

For the lower segment

$$p = \frac{W}{Hh_2^2}(2h_1 + h_2)(w - y) \tag{11.7}$$

Eliminating p from Equations 11.6 and 11.7 gives

$$2h_1 = h_2 \tag{11.8}$$

Since $h_1 + h_2 = H$, $h_1 = \frac{H}{3}$ and Equation 11.7 becomes

$$p = \frac{3W}{H^2}(w - y) \tag{11.9}$$

Since the pressure is synonymous with the unit resistance, Equation 11.9 provides phase (ii) of the resistance function. Phase (i) is provided by simple elastic analysis

$$r = \frac{384EI}{5H^4} \cdot y$$

The transition from phase (i) to phase (ii) occurs when the masonry cracks. This occurs when

$$r = \frac{4}{3}\left(\frac{w}{H}\right)^2 \cdot \sigma_m$$

where σ_m is the tensile strength of mortar. The final form of the resistance function is shown in Figure 11.10.

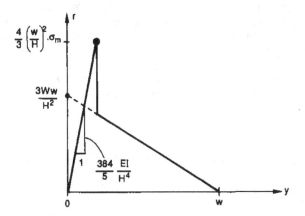

Figure 11.10 *Resistance function for masonry wall*

11.4.4 Arching effect in masonry walls

If vertical movement of the upper edge of the slab is prevented, the wall can only fail if crushing of the masonry takes place along the top and bottom edge of the wall and along the failure surface. Theoretical analyses indicate that this effect, known as *arching*, can enhance the resistance of the wall by a factor of 50 or more. However, such figures are not to be relied upon since only very slight in-plane movements of the peripheral supports will negate the predicted benefits.

If a masonry structure is to be used to provide blast protection, which is in itself inadvisable, the resistance can be maximised by incorporating arching. However, it would be foolhardy to rely on the theoretically predicted improvement which the arching effect provides.

11.5 Methods of solution

There are many methods available to allow solution of the equations of motion that describe the behaviour of blast loaded structures. It is sometimes possible to solve these equations exactly as described in Chapter 8. Generally, though, a numerical solution usually involving the use of a computer is necessary and there are many texts that present solution techniques[3,4,5]. Here mention will be made of a method that can yield acceptable answers by hand calculation.

The equation that we wish to solve is 11.1 recast below as

$$M\ddot{x} + R(x) = F(t) \tag{11.10}$$

where $R(x)$ is the resistance function for the structure. To evaluate the response of the structure we will use the 'predictor' method of solution in which the response of the target is considered in small time steps Δt during which period the acceleration of the target is considered constant.

To illustrate the method, consider a resistance term $R(x)$ which has the form shown in graph Figure 11.11a while the blast load variation with time is given in Figure 11.11b. At the start of the calculation ($t = 0$) the displacement x and the velocity x of the target are zero and so, right at the beginning, the differential equation can be rearranged to give the initial acceleration

$$\ddot{x}_0 = \frac{F(0) - R(x_0)}{M} \tag{11.11}$$

which is the first value shown in Figure 11.11c. The velocity at the end of this time step is

$$\dot{x}_1 = \dot{x}_0 + \ddot{x}_0 \Delta t \tag{11.12}$$

and the displacement is given by

$$x_1 = \dot{x}_0 + \dot{x}_0 \Delta t + \tfrac{1}{2}\ddot{x}_0 \Delta t^2 \tag{11.13}$$

From this displacement the value of R can be read from Figure 11.11a and, by reading the value of F at the start of the second time step, x_1 can be found and so on. This process can be continued until the complete displacement response is evaluated as shown in Figure 11.11d. To illustrate the technique, consider a specific example.

Example 11.1
In order to provide restraint against translatory motion under the action of a blast load, a small radar installation of mass 500 kg is tethered by a cable to the ground. An idealisation of the system is shown in Figure 11.12a. The cable can stretch by 20 mm before it breaks. Its load-extension characteristics are shown in Figure 11.12b.

The installation is subject to a blast wave whose peak pressure corresponds to a load of 4000 N which then decays linearly as given by

$$F(t) = 4000(1 - t/0.1) \tag{11.14}$$

Show how the acceleration, velocity and displacement vary with time and say if the cable is sufficiently strong.

The solution is best carried out in tabular form using the calculation grid of Table 11.1 (p. 222). Solution commences with the evaluation of the initial acceleration using Equation 11.11 as

$$\ddot{x}_0 = \frac{4000 - 0}{500} = 8\,\text{m/s}^2 \tag{11.15}$$

From this, using Equations 11.12 and 11.13, velocity and displacement can be evaluated at the end of the first time step. The resulting displacement can then be used to find R_1, the resistance at the start of the next time step and hence the associated acceleration, velocity and displacement and so on. The grid shows the results of such calculations.

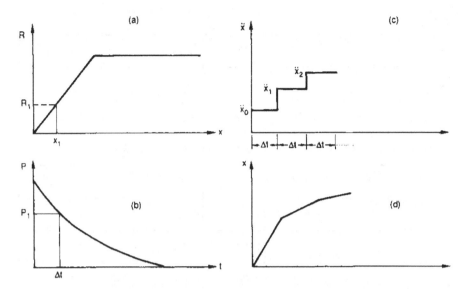

Figure 11.11 Graphical representation of numerical solution of equation of motion

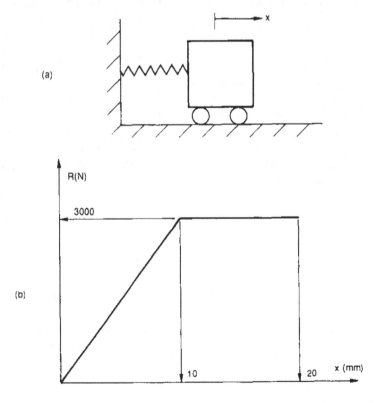

Figure 11.12 SDOF system and associated resistance-deflection graph for Example 11.1

Table 11.1

n	t(s)	$R_n(N)$	$P_n(N)$	$\ddot{x}_n(m/s^2)$	$\dot{x}_n$ (m/s)	x_n (mm)
0	0	0	4000	8.000	0	0
1	0.01	120.0	3600	6.960	0.080	0.400
2	0.02	464.4	3200	5.471	0.150	1.548
3	0.03	996.5	2800	3.607	0.205	3.322
4	0.04	1665.7	2400	1.469	0.241	5.552
5	0.05	2410.7	2000	−0.821	0.256	8.035
6	0.06	3000.0	1600	−2.800	0.248	10.554
7	0.06	3000.0	1200	−3.600	0.220	12.894
8	0.08	3000.0	800	−4.400	0.184	14.914
9	0.09	3000.0	400	−5.200	0.140	16.534
10	0.10	3000.0	0	−6.000	0.088	17.674
11	0.11	3000.0	0	−6.000	0.028	18.254
12	0.12	3000.0	0	−6.000	−0.032	18.234
13	0.13	3000.0	0	−6.000	−0.092	17.614

The algorithms used to calculate acceleration, velocity and displacement are:

$$\ddot{x}_n = \frac{P_n - R_n}{M} : \dot{x}_n = \dot{x}_{n-1} + \ddot{x}_{n-1}\Delta t : x_n = x_{n-1} + \dot{x}_{n-1}\Delta t + \tfrac{1}{2}\ddot{x}_{n-1}\Delta t^2$$

The calculations show that the maximum extension of the cable is just under 18.3 mm, indicating that the cable will not break. The variation of acceleration, velocity and displacement with time is shown in Figure 11.13.

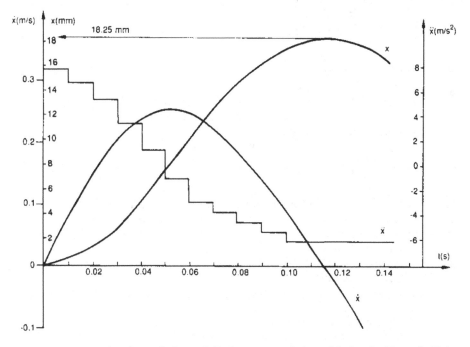

Figure 11.13 *Acceleration, velocity and displacement variation with time for Example 11.1*

11.6 References

[1] TM5-1300: *Design of Structures to Resist the Effects of Accidental Explosions.* US Department of the Army (1991)

[2] Keating J., Hetherington J. G., Mays G. C., Load-Deformation Characteristics for Wall Panels with Openings. Final Report on Agreement No. 2004/118/RARDE, September 1993.

[3] Griffiths D.V., Smith I.M., *Numerical Methods for Engineers.* Blackwell Scientific Publishers, Oxford (1991)

[4] Fagan M.J., *Finite Element Analysis: Theory and Practice.* Longman Scientific and Technical, London (1992)

[5] Mason J.C., *BASIC Numerical Methods.* Butterworth Scientific, London (1983)

Symbols

A	moment arm length
B	moment arm length
C	moment arm length
E	Young's modulus
$f_1(x)$	structure resistance
$f_2(x)$	structure resistance

$F(t)$ blast load
$F_x(t)$ blast load in x direction
$F_z(t)$ blast load in z direction
F_c cable force
F_F structure resistance (e.g. friction)
F_I inertial resistance
g gravitational acceleration
h_1 distance from top of masonry wall to fracture
h_2 height of fracture of masonry wall above ground
H height of masonry wall
I second moment of area of target structure
K cable stiffness
L span of slab
M vehicle mass
M_N ultimate negative moment capacity per metre run of slab
M_p ultimate positive moment capacity per metre run of slab
p uniform loading pressure
r resistance of masonry wall
r_e pressure to cause yield
r_u ultimate resistance of slab
$R(x)$ generalised structure resistance
t time
w thickness of masonry wall
W weight per metre run of masonry wall
x linear displacement
x_e deflection at yield
x_o initial deflection at $t = 0$
x_p maximum deflection of slab
y displacement of inner fracture surface of masonry wall
z linear displacement
θ rotational displacement
μ coefficient of friction
σ_m tensile strength of mortar

12 Protection against ballistic attack

12.1 Introduction

Evolution of the ballistic threat has taken place, principally, in the contexts of general war, terrorism and crime. The threat has been developed to cause injury or damage to personnel, land vehicles, ships, aircraft and structures. Inert projectiles will cause only localised damage to structural targets and therefore normally constitute a less potent threat to the survival of the structure than rounds with high explosive content. Nonetheless, provision of adequate protection to personnel and equipment within structural targets against projectiles and fragments can be of crucial importance in protective design. This chapter describes the spectrum of ballistic threats to which a target could be subjected and the mechanisms by which each threat causes damage to the target. A discussion of terminology used in describing projectile-target interaction is presented, together with the definitions of fundamental parameters used in the study of terminal ballistics. A range of penetration prediction equations is presented and their applicability to the design of armour systems is discussed. This is complemented by a brief description of the way in which computer codes can be used in predicting target response. Finally, some modern armour strategies, normally adopted for mobile targets, will be discussed with particular reference to their applicability for structural targets.

12.2 The ballistic threat

For convenience, a ballistic threat is normally categorised as either a *kinetic energy* or a *chemical energy* threat. A kinetic energy threat is one in which penetration is achieved by an inert projectile, by virtue of the fact that it possesses kinetic energy. The kinetic energy will have been given to the projectile either at launch, as in a gun, or during flight, as in the case of a rocket-powered missile system. Examples of kinetic energy threats include most small-to-medium calibre rounds, larger calibre armour piercing (AP) rounds and, the most potent threat of all, the long rod penetrator.

A chemical energy threat is one in which the energy is delivered to the target as stored chemical energy in the form of high explosive. Examples of the chemical energy threat include shaped-charge rounds, and high explosive squash head (HESH) rounds. In the shaped-charge round (Figure 12.1), the explosive is detonated by a fuzing system just prior to impact, forming

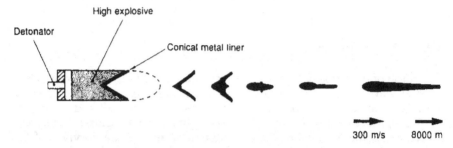

Figure 12.1   *Schematic representation of a shaped-charge round*

the conical metal liner into a highly penetrative jet. In the HESH round a 'pat' of explosive is detonated on the face of the target shortly after impact, causing damage in the target by a combination of blast waves and stress waves (see Chapter 6). Although the HESH round is normally fired from a tank gun, shaped charges find application in tank ammunition, missile systems, rocket-propelled grenades (RPGs), mines and demolition devices.

Whether a fragmentation shell is a chemical energy or a kinetic energy threat is a debatable point. Although the effect of the system relies upon the delivery of a quantity of explosive to within close proximity of a target, the penetration of the target is achieved by the kinetic energy of inert fragments. For this reason it is more helpful to categorise the threat in terms of the mechanism by which damage is caused to the target.

12.3   Impact regimes

A key parameter in determining target response to an impact is the *kinetic energy density* delivered by the projectile. This is defined as the kinetic energy of the projectile divided by its cross-sectional area. When the kinetic energy density at the impact site is low, the shear stress generated in the target may be of the same order of magnitude as the shear strength of the target. Conventional 'strength of materials' parameters (e.g. strength, stiffness, hardness and toughness) will determine the penetration mechanism and will govern the extent of the penetration. This is known as the *subhydrodynamic regime* of penetration. For longer, thinner, heavier and faster projectiles (e.g. long rod penetrators) the kinetic energy density will be greatly increased. The shear stresses generated on impact may be many orders of magnitude greater than the shear strength of both the target and the penetrator. In this case the strengths of the materials are negligible and the impact can be characterised as a fluid–fluid interaction, governed by the laws of fluid dynamics. This is known as the *hydrodynamic regime* of penetration. The transition between the subhydrodynamic regime and the hydrodynamic regime extends over quite a wide velocity range. Below 1000 m/s all impacts are subhydrodynamic, above 3000 m/s all impacts are hydrodynamic. In the transition zone, although fluid flow governs the interaction with density being the dominant physical parameter, strength still proves to be a significant factor. Long rod penetrators, with impact velocities in the region of 1600 m/s, are clearly in the transition zone whereas shaped-charge penetration, with jet tip velocities in excess of 8000 m/s, is purely hydro-dynamic.

12.4 Stress waves, scabbing and spalling

When a projectile strikes a target, transient stresses are generated which then propagate away from the impact site. The shear stress component of the transient stress pulse may cause a conical fracture surface emanating from the point of impact and the compressive stress pulse will reflect from the rear face of the target as a tensile pulse which may cause scabbing. Whether or not a scab is formed depends on the stress levels induced by the impact and the fracture toughness of the target. Such transient stresses are also gener-ated in targets by explosives detonated on the face of a target, as in the HESH round, and, to a lesser extent, by blast waves. In all of these cases, damage can occur to personnel and equipment, caused by flying debris from the target, even though the target has not been perforated by the threat. This topic is dealt with in greater detail in Chapter 6.

12.5 Penetration performance parameters

Terminal ballistics events are frequently described using a number of com-mon terms and parameters. Confusion can arise when parties assume dif-ferent meanings for these parameters and so it is always good practice to quote definitions of any parameter used. In this section some important quantities used in terminal ballistics are defined and discussed.

12.5.1 Penetration and perforation

The term *perforation* implies that a projectile has passed through a target whereas *penetration* is defined as the distance the tip of a projectile has travelled into a target. Thus perforation is synonymous with complete pene-tration. This general definition is open to a number of interpretations, as can be seen from Figure 12.2.

By the US Army definition perforation is achieved by a round when it can be seen on the rear face of the target. By the US Navy definition, also adopted in the UK, perforation is achieved only when the round has passed completely through and emerged from the target. The US Protection Limit requires that only some part of the projectile or the target perforates a thin witness screen placed 150 mm behind the rear face of the target. These small differences in definition can be sufficient to win or lose a contract or, more significantly, a matter of life or death. It is therefore crucial when quoting required protection levels, or interpreting a specification, to be unambiguous in the definition of perforation.

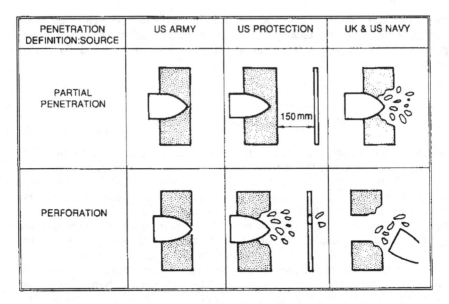

PENETRATION DEFINITION:SOURCE	US ARMY	US PROTECTION	UK & US NAVY
PARTIAL PENETRATION		150 mm	
PERFORATION			

Figure 12.2　*Definition of partial penetration and perforation*

12.5.2　Ballistic limit velocity

The ballistic limit velocity is, essentially, the velocity at which a projectile must be travelling in order to perforate a specified target. It is assumed that the projectile strikes the target with zero yaw and zero obliquity (see Section 12.5.3). In reality, despite meticulous efforts to replicate experimental conditions, a ballistic event repeated several times can yield a number of different results. For this reason, the ballistic limit velocity of a specified penetrator is defined as the velocity at which the projectile, striking a specified target with zero yaw and obliquity, has a 50% chance of perforating the target. Evaluating the ballistic limit velocity, or V_{50}, experimentally for a particular projectile-target combination can prove time-consuming and expensive. A 'short cut' procedure is therefore normally adopted in which the velocity of the round is varied to achieve three complete penetrations (i.e. perforations) and three partial penetrations within a small velocity band. The width of this velocity band varies from specification to specification, but is typically 36 m/s. The six velocities are then averaged to give a value for V_{50}.

12.5.3　Obliquity and yaw

Although the definitions of penetration, perforation and ballistic limit have been based on the concept of a round striking a target normally, in reality this almost never happens. Figure 12.3 shows a round travelling towards a target with both obliquity and yaw. The obliquity θ is defined as the angle between the round's trajectory and the normal to the target. The round's yaw ψ is defined as the angle between the axis of symmetry of the round and the

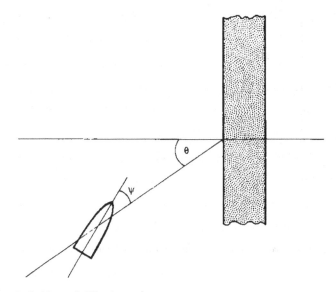

Figure 12.3 *Definitions of obliquity and yaw*

tangent to its trajectory. Both obliquity and yaw normally degrade the performance of a projectile, a fact which can be used to advantage by the armour designer. This topic will be discussed in more detail in Sections 12.5.4 and 12.8.

12.5.4 Equivalent protection factors

Experimental observations show that inclining an armour plate with respect to the threat trajectory improves its effectiveness. Diagrams a and b in Figure 12.4 show the same plate arranged in two different orientations with respect to the threat. The inclined plate, provides superior protection to the vertical plate for two reasons. Firstly, and most significantly, there is more material placed in the path of the round. This is referred to as the *geometric* effect. Secondly, because the round strikes an inclined surface, its path is disrupted and it follows a less energetically efficient path through the target. This is referred to as the *disruptive* effect. The combined action of the geometric and disruptive effects means that thinner armour plates can be used to provide equivalent protection if they are inclined. Diagram c shows an inclined plate which has been shown in trials to provide the same level of protection as the plate in a. The degree of thickness reduction possible is described by the *equivalent protection factor* (EPF) which is defined as follows:

$$EPF = \frac{\text{thickness of armour required to defeat the threat at obliquity } \theta}{\text{thickness of armour required to defeat the threat at normal}}$$

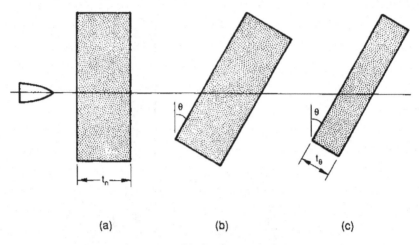

Figure 12.4 *Improved effectiveness of inclined armour plates*

i.e.

$$EPF = \frac{t_\theta}{t_n}$$

Trials information on the performance of a specific round/target-material combination is commonly displayed in graphical form (Figure 12.5). The argument which is normally put forward in favour of inclining armour is one of potential weight saving. Weight saving may not be important to the civil engineer and may prove illusory to the vehicle designer. The desirability of inclined armours is discussed further in Section 12.8.

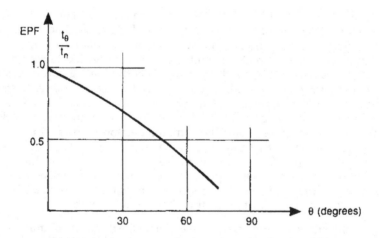

Figure 12.5 *Typical EPF graph*

12.5.5 Areal density

For mobile targets it is important to provide protection at minimum weight. Armour technologists seek to develop protective systems which are both effective and lightweight. The parameter which is used to describe the weight of an armour system is its *areal density*. The areal density of an armour system is the mass (in kg) of a 1 metre square slab of the armour.

Thus, areal density $= \rho \times t$ where ρ is the average bulk density of the armour system (kg/m^3) and t is thickness of the armour (m).

Example 12.1
Trials show that a 5.56 mm SS109 round will be stopped by a 6 mm plate of rolled homogeneous steel armour (RHA) or by a 14 mm plate of aluminium armour. Compare the areal densities of the two systems and discuss the implications of the result. ($\rho_{steel} = 7850$ kg/m^3, $\rho_{Al} = 2800$ kg/m^3).

Solution
Areal density of RHA plate to defeat 5.56 round
$$= 0.006 \times 7850 = 47.1 \text{ kg/m}^2$$
Areal density of aluminium plate to defeat 5.56 round
$$= 0.014 \times 2800 = 39.2 \text{ kg/m}^2$$

These figures indicate that a 17% weight saving can be achieved by adopting aluminium in place of steel, making aluminium the preferred choice. However, there are many other important considerations which may influence material selection. Firstly, the threat considered here is relatively low and only modest thicknesses of armour are required to provide protection. The extra bulk of the aluminium armour can probably be accommodated. At higher threat levels, though, bulk can be the deciding factor. A long rod penetrator, for example, can currently penetrate about 600 mm of RHA or over 1000 mm of aluminium armour. In broad terms, aluminium is a candidate material for armoured personnel carriers (APCs) and reconnaissance vehicles but can be ruled out for main battle tanks (MBTs) on the grounds of excessive bulk. Interestingly, a comparison of an aluminium plate and a steel plate chosen to provide equal protection reveals that the aluminium plate is the stiffer, even though steel is a stiffer material than aluminium. This means that, for vehicles designed to provide equal protection, the aluminium vehicle will be inherently more rigid. This fact can lead to an extra weight saving for aluminium vehicles by using less substantial structural supporting members.

Other arguments used against selecting aluminium include:

(a) welding difficulties, particularly in field repair;
(b) greater susceptibility to stress corrosion cracking;
(c) enhanced behind armour effects: pyrophoricity, spalling, overpressures, etc;
(d) cost.

Considerable progress has been made in recent years in overcoming these drawbacks and aluminium is currently the preferred choice for vehicles below about 30 tonnes.

12.6 Penetration prediction equations

The aim of a penetration prediction equation is to provide an estimate of the distance a penetrator will travel into a target. This enables the designer to specify appropriate armour levels to offer protection against the perceived threat. Many prediction equations employ specialised materials data, often obtained from ballistic trials or high-strain-rate experiments. Acquiring such data often causes difficulty for the designer, particularly when assessing the potential performance of novel materials. For this reason the prediction methods presented in the following section draw only on material data which should be readily available. Generalised prediction equations are presented for the subhydrodynamic and hydrodynamic regimes with specialist sections on reinforced concrete, soils and composite armours.

12.6.1 Subhydrodynamic regime

R. F. Recht[1] developed a comprehensive penetration prediction equation which works well for a wide range of target materials in the subhydrodynamic regime (see Section 12.3). It is relevant for small to medium calibre rounds, fragments and all projectiles with a length to diameter ($^L/_D$) ratio of less than 10 with impact velocities less than 1000 m/s. Rounds with a high $^L/_D$ ratio, which are normally sabot-mounted, often induce hydrodynamic response and will be dealt with in Section 12.6.2. In the development of his analysis, Recht examines a *rigid* projectile penetrating a *semi-infinite, ductile* target. Although these conditions are seldom met, the Recht equation can still provide a very useful estimate of penetration. The equation can be presented in the following form:

$$\int_0^x \left(\frac{A_x}{A_p}\right) dx = \frac{1.61M}{bA_p}\left[(V_0 - V) - \frac{a}{b}\ln\left(\frac{a + bV_0}{a + bV}\right)\right] \tag{12.1}$$

where
$$a = 2\tau \ln(2z)\left(1 + \frac{f}{\tan \alpha}\right)$$

and
$$b = 0.25\sqrt{K\rho}\left(1 + \frac{f}{\tan \alpha}\right)\sin \alpha$$

and A_x and A_p are the cross-sectional areas of the projectile as defined in Figure 12.6.

Since the projectile is assumed to remain rigid, all of the material properties $\left(\tau, K, \rho \text{ and } z, \text{ where } z = \left(\frac{E}{\sigma_y}\right)\left(1 + \frac{2E}{\sigma_y}\right)^{-1/2}\right)$ relate to the target. The parameter f is the dynamic coefficient of friction between round and target and is normally very small (about 0.01) for metal-metal interfaces. Consequently the term $\left(1 + \frac{f}{\tan \alpha}\right)$ can normally be set to unity without significant loss of accuracy. M is the mass of the projectile whilst α represents the half-angle of

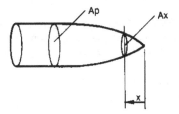

Figure 12.6   *Projectile geometry*

a conically nosed projectile. Recht showed that, for projectiles with standard ogives, a good estimate is achieved by setting $\alpha = 23.5°$. For a cylindrical penetrator, $A_x = A_p$ along the length of the penetrator and the left hand side of Equation 12.1 reduces to x, the penetration. We can therefore think of the left hand side of the equation being equal to the penetration, modified by a geometric factor. For rough estimates it may be acceptable to take the geometric factor to be unity, which will introduce an error of less than 10%. The equation can provide an estimate of the penetration at any stage, since V_0 represents the impact velocity and V the current velocity. By setting V to a range of values between V_0 and 0, a velocity-penetration history, such as that shown in Figure 12.7, can be developed. The Recht equation has been shown to provide good estimates of penetration for a wide range of materials including plastics, ceramics, metals and concrete. Moreover, with its incremental treatment of the variation of projectile presented area, it provides a useful method for analysing the penetration caused by fragments.

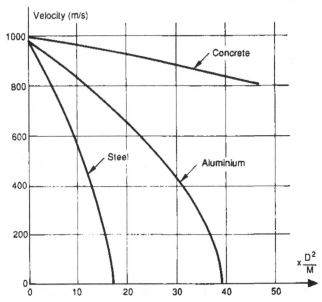

Figure 12.7   *Predictions of velocity-penetration histories from the Recht equation*

Example 12.2

A cylindrical steel penetrator of diameter 10 mm has a shaft of length 50 mm and a conical nose of length 9 mm. Estimate the penetration which will be achieved when it strikes an aluminium alloy target at 900 m/s. Take the following data for the target:

$$K = 6.9 \times 10^{10} \text{N/m}^2$$
$$E = 7.1 \times 10^{10} \text{ N/m}^2$$
$$\sigma_y = 3.4 \times 10^8 \text{N/m}^2$$
$$\tau = 2.8 \times 10^8 \text{N/m}^2$$
$$\rho = 2800 \text{ kg/m}^3$$
and $f = 0.01$

Solution

$$\text{Volume of projectile} = \frac{\pi 10^2}{4}\left(50 + \frac{9}{3}\right) = 4.163 \times 10^3 \text{mm}^3$$

$\therefore$ Mass of projectile $= 4.163 \times 7.85 = 32.7$g.

The penetration prediction has to be conducted in two stages:

1 To determine the residual velocity of the penetrator when the nose is just totally embedded in the target.
2 To determine the subsequent penetration achieved.

When the tip of the nose has travelled a distance x, the presented area A_x is

$$\frac{\pi}{4} \cdot \left(\frac{x}{9} \cdot 10\right)^2$$

So at this stage $\dfrac{A_x}{A_p} = \dfrac{x^2}{81}$

and the LHS of equation of 12.1 becomes:

$$\int_0^9 \frac{x^2}{81} dx = \left[\frac{x^3}{243}\right]_0^9 = 3\text{mm} = 3 \times 10^{-3}\text{m}$$

Equation 12.1, with $a = 1.7 \times 10^9$ N/m^2 $b = 1.72 \times 10^6$ Ns/m^3 $V_0 = 900$ m/s becomes

$$3 \times 10^{-3} = 3.9 \times 10^{-4}\left(900 - V - 988\ln\left(\frac{1888}{988 + V}\right)\right)$$

Solution by trial and error gives $V = 884$ m/s, which is the residual velocity of the penetrator when the tip is completely embedded in the target. During the second phase of the calculation, $A_x = A_p$ and the left side of the equation reduces to x, the penetration. Thus setting $V_0 = 884$ and $V = 0$ gives $x = 3.9 \times 10^{-4}(884 - 988\ln 1.895) = 98.6$ mm. So total penetration $= 9 + 98.6 = 107.6$ mm.

12.6.2 *Hydrodynamic regime*

The first order hydrodynamic model is based on the assumption that the interaction between projectile and target can be described by a simple fluid dynamics model of behaviour. Penetration is achieved by a process of mutual erosion of the projectile and target, as shown in Figure 12.8. Since the crater base is travelling at speed u whilst the rod arrives at speed v, the rod is being consumed at a rate of $v - u$ m/s. For a rod of length L the duration of the event is $\frac{L}{v-u}$ seconds, during which time the base of the crater advances a distance

$$\frac{u \cdot L}{v - u} \tag{12.2}$$

A consideration of the pressure at the point O leads to

$$\rho_p(v - u)^2 = \rho_t u^2$$

and therefore

$$\frac{u}{v - u} = \sqrt{\frac{\rho_p}{\rho_t}} \tag{12.3}$$

Combining Equations 12.2 and 12.3 leads to

$$\text{penetration} = L\sqrt{\frac{\rho_p}{\rho_t}} \tag{12.4}$$

which is known as the first order hydrodynamic equation.

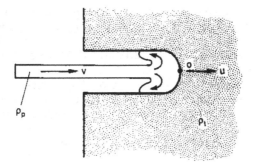

Figure 12.8 *Schematic representation of long rod penetration*

Since the derivation of Equation 12.4 assumes pure hydrodynamic behaviour, material strength plays no part. The only significant material property is the density and the penetration is independent of impact velocity. Experimental data obtained by Hohler and Stilp[2] for tungsten alloy rods fired at a range of velocities at steel targets is shown in Figure 12.9. The ordinate represents the measured penetration divided by the penetration predicted using Equation 12.4. At velocities in excess of 3000 m/s true hydrodynamic behaviour prevails and predictions from Equation 12.4 conform with experimental data. Current long rod impact velocities, which are in the region of 1600 m/s, are not sufficiently high to generate true hydrodynamic behaviour, with the result that the penetration achieved is only about 65% of that predicted by the first order model.

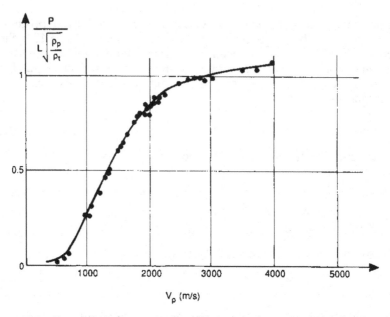

Figure 12.9   *Penetration of tungsten alloy rods into steel targets (after Ref. 2)*

Due to the very high velocities involved, Equation 12.4 is successful in predicting the penetration of shaped-charge rounds. However, due to the difficulty in determining the length of the jet, it is more common to use the empirical data presented in Figures 12.10 and 12.11 for conventional copper jets penetrating steel targets.

The diagrams are applicable to any size of shaped charge since distances are presented in cone diameters (i.e. the diameter of the conical liner in the shaped-charge round). The figures convey the importance of cone angle and distance from the target at detonation (i.e. stand-off) to the resultant penetration. Penetration into other target media can be estimated by invoking Equation 12.4 as follows:

$$P = P_{st}\sqrt{\frac{\rho_{st}}{\rho}} \tag{12.5}$$

where P_{st} is the penetration deduced from Figures 12.10 and 12.11 ρ_{st} is the density of steel and ρ is the density of the chosen target material.

Although Equation 12.4 overestimates the penetration which will result from a long rod penetrator, it does provide some assistance in developing a protection strategy. If the bulk of a protective system is critical, it is important to keep penetration to a minimum. This is achieved by selecting an armour material with a high density, since penetration is inversely proportional to the square root of the target density. However, if weight is the dominant criterion, low density materials should be chosen, since areal density is directly proportional to the square root of the density.

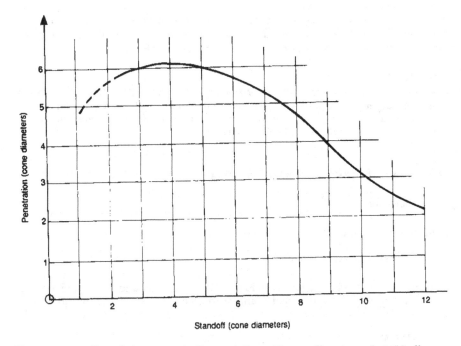

Figure 12.10 *Shaped-charge penetration variation with cone diameter and stand-off*

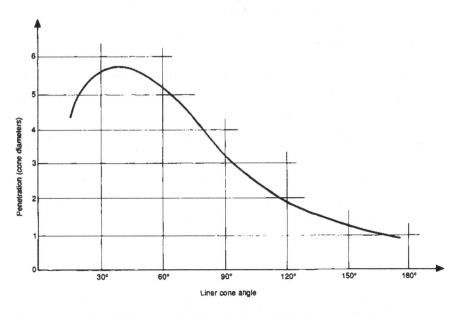

Figure 12.11 *Shaped-charge penetration variation with liner cone angle*

In recent years the effectiveness of the shaped-charge round on the battle-field has been greatly reduced by the emergence of explosive reactive armour. This will be discussed further in Section 12.8.

Example 12.3
A steel long rod penetrator, of diameter 26 mm and $^L/_D$ ratio 24, is fired at 1600 m/s at a steel target. Estimate the penetration that would result. What would be the penetration achieved by a rod of identical geometry made of depleted uranium? (Take $\rho_{st} = 7850$ kg/m^3, $\rho_{DU} = 18\,600$ kg/m^3).

Solution
The first order hydrodynamic model, Equation 12.4, gives

$$\text{penetration} = L\sqrt{\frac{\rho_p}{\rho_t}}$$

but $L = {^L}/_D \cdot D = 24 \times 0.026 = 0.624$ m.

For steel penetrator and steel rod $\dfrac{\rho_p}{\rho_t} = 1$

So penetration = 0.624 m. Referring to Figure 12.9, true penetration will be 65% of this value, i.e. 0.406 m. For a depleted uranium rod, penetration will be increased by a factor

$$\sqrt{\frac{18\,600}{7850}}$$

$\therefore$ penetration of DU penetrator = 0.625 m.
Note: it is a good rule of thumb that a heavy metal rod (i.e. tungsten alloy or depleted uranium) will achieve a penetration into a steel target approximately equal to its own length at an impact velocity of about 1600 m/s.

12.6.3   Reinforced concrete

Although the techniques described in the previous sections provide adequate predictions of penetration into concrete, a more detailed treatment is warranted due to its importance as a protective medium for structures. A significant amount of research effort has been devoted to studying the ballistic penetration of concrete, resulting in a large number of empirical equations. Military Engineering Volume IX[3] recommends the use of the US NDRC formula as modified by Barr *et al*[4]. Figure 12.12 shows the penetration which would be achieved by a projectile into a semi-infinite mass of reinforced concrete, whereas Figure 12.13 shows the thickness of concrete slab which will just be perforated by a specified threat. A significant hazard can result from scabbing in concrete targets. Figure 12.14 gives recommended slab thicknesses to eliminate the risk of scabbing.

12.6.4 Soils

Granular materials comprise three phases – solids, liquid and air – and are often inhomogeneous and anisotropic. They therefore present the designer

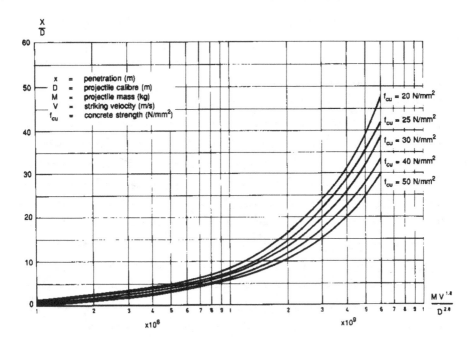

Figure 12.12 *Projectile penetration into reinforced concrete (after Ref. 3)*

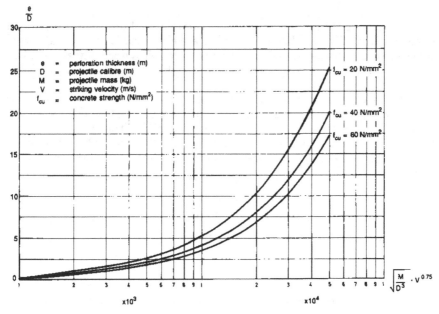

Figure 12.13 *Thickness of concrete slabs to resist perforation (after Ref. 3)*

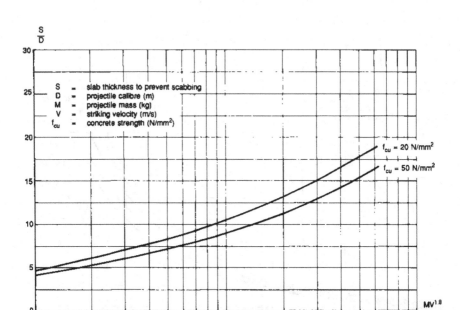

Figure 12.14 *Recommended slab thickness to eliminate scabbing (after Ref. 3)*

with a significantly more complex and variable medium than other target materials. Inevitably prediction equations are empirical in nature and can be subject to quite large errors. Characterisation of soil is normally achieved by identifying a number of discrete soil types to which a value of a soil constant is assigned.

Research workers at the Sandia National Laboratories[5] have developed prediction equations for normal impact of long penetrators ($^L/_D > 10$), for which the penetration paths are essentially straight. The influence of soil type is effected through the use of a soil constant S, the value of which is small for hard soils and large for soft soils. Some example values are given in Table 12.1. The value of the projectile nose performance coefficient N is given by

$$N = 0.56 + 0.183\,n \quad \text{for ogival noses}$$

and $\quad N = 0.56 + 0.25\,n \quad \text{for conical noses}$

where $n = \dfrac{\text{nose length}}{\text{round diameter}}$

Table 12.1

Soil type	*S*
Sandia equation – soil constant 'S'	
Rock	1 – 2
Frozen soils, glacier ice	2 – 4
Hard, dry clay	4 – 6
Densely packed sand	4 – 6
Loosely packed sand	6 – 8
Stiff clay	8 – 12
Very loosely packed sand	8 – 12
Topsoil	10 – 15
Medium stiff clay	10 – 15
Very loose topsoil with humus	20 – 30
Very weak wet clay or mud	40 – 50

The equations are as follows:

1 For impact velocities less than 61 m/s

$$\text{penetration(m)} = 6.06 \times 10^{-3} . S . N . \left(\frac{M}{A}\right)^{1/2} . \ln\left(1 + 2.15 \times 10^{-4} V_0^2\right)$$

$$(12.6)$$

2 For impact velocities greater than 61 m/s

$$\text{penetration (m)} = 1.16 \times 10^{-4} . S . N . \left(\frac{M}{A}\right)^{1/2} (V_0 - 30.5) \qquad (12.7)$$

where M is penetrator mass (kg), A is cross-sectional area of penetrator (m²) and V_0 is impact velocity (m/s).

The Sandia equations will give predictions with an error of less than 20% provided the following conditions are met:

1 The mass of the projectile lies in the range 27–2600 kg.
2 The slenderness ratio, L/D, is greater than 10.
3 The predicted penetration exceeds three penetrator body diameters plus one nose length.

For less slender projectiles $(L/D < 10)$, the US National Defense Research Committee (NDRC)[6] method can be used.

Due to the instability of the penetrator, the projectile's path in the soil is often curved and so, initially, the total path length in metres L (Figure 12.15) is estimated as follows:

$$L = M^{1/3}\left(A + B\sqrt{V_0 - C}\right) \qquad (12.8)$$

where A, B and C are given for various soil types and projectile nose geometries in Table 12.2.

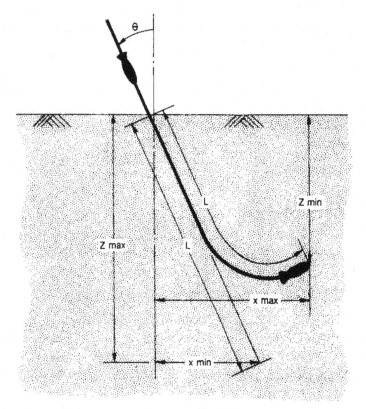

Figure 12.15   *Curved trajectory in soil*

Table 12.2

	Blunt nose (R/d = 0.5)†			Average nose (R/d = 1.5)†			Sharp nose (R/d = 3)†		
Soil type	A	B	C	A	B	C	A	B	C
Sand	0.451	0.0301	128.8	0.451	0.0301	128.8	0.451	0.0301	128.8
Sandy loam	0.508	0.0418	135.8	0.493	0.0397	134.7	0.493	0.0397	134.7
Loam	0.640	0.0503	152.5	0.554	0.0488	135.4	0.797	0.0350	111.6
Clay	0.763	0.0722	113.3	1.022	0.0528	138.7	0.972	0.0515	160.2

†R/d =Radius of ogive/projectile diameter.

For a projectile striking the ground at an obliquity θ (Figure 12.15) the maximum vertical depth to which the penetrator will reach is $L\cos\theta$ and it will travel at least a distance $L\sin\theta$ horizontally. For a projectile with a curved trajectory, the vertical depth of penetration may be as little as $\frac{2}{3}L\cos\theta$, whereas the horizontal travel may increase to $\frac{2}{3}L\sin\theta + \frac{L}{3}$.

For situations not covered by the foregoing methods, resort may be made to the subhydrodynamic and hydrodynamic models presented earlier in this chapter.

12.6.5 Composite systems

Protection may be provided to a potential target by more than one layer of protective material. In the case of a buried structure, for example, a burster slab may overlie one or more strata of soil, covering the concrete structure itself. In such cases, it is assumed that each component in the protective system acts independently and the total protection offered is simply the aggregate of that offered by each element. The predictive methods presented earlier are used to evaluate the residual velocity of the penetrator as it emerges from the first layer which is then used as the impact velocity for the second layer and so on.

There are, however, some armour systems, known as composite, laminated and complex armours, in which the composite action of the constituent elements exceeds the sum of the contributions made by each component. A common form of composite armour is lightweight, ceramic-faced armour which currently finds application in body armour, internal security, light vehicles, reconnaissance vehicles and light tanks. Such systems are being considered for protection of static emplacements and sensitive areas in structural installations. These armours comprise a ceramic front face affixed to a ductile backing plate. Typically, aluminium oxide or boron carbide will be used for the ceramic front plate and steel, aluminium, glassfibre reinforced plastic or an aramid fibre composite for the backing plate. The action of the composite system is illustrated in Figure 12.16. The hard ceramic blunts the penetrator, while a system of stress waves, propagating through the plate, causes the ceramic to shatter into a conoid of rubble. This process spreads out the kinetic energy of the round so that it can now be coped with by the backing plate.

The backing plate material has been chosen for its ability to deform extensively before breaking, absorbing the kinetic energy of the round and the ceramic rubble as it does so. The composite action described above means

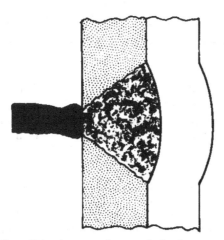

Figure 12.16 *Spreading of kinetic energy in ceramic-faced armours*

that ceramic-faced armours can provide a 40% weight saving over RHA. Their disadvantages are that they are not ideal structural materials and are therefore normally used as appliqué systems and, due to the fracture properties of ceramics, their multi-hit capability is inferior to that of metallic armours.

Due to the complex action of ceramic-faced armours, performance prediction is difficult. A method developed by Florence[7] for armour piercing (AP) rounds striking alumina-aluminium targets is recommended for its effectiveness and simplicity.

Florence made the empirical observation (Figure 12.17) that the radius of the base of the ceramic fracture conoid was given by $(a_p + 2h_1)$, where a_p is the radius of the penetrator and h_1 is the thickness of the ceramic front plate. He then assumed that the back plate deformed as a circular membrane or

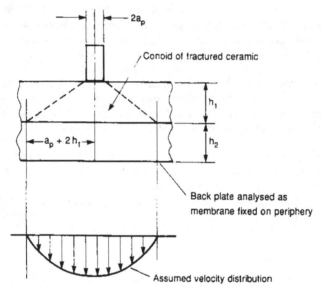

Figure 12.17   *Florence's assumption of conoid geometry*

radius $(a_p + 2h_1)$, pinned around its periphery. The kinetic energy of the round and ceramic is equated to the work done in stretching the membrane until it reaches its tensile breaking strain. By this method the following expression is derived for the ballistic limit velocity of the round.

$$V_{50} = \left(\frac{\epsilon\, S}{0.91 \cdot M \cdot f(a)} \right)^{\frac{1}{2}}$$

Where M is projectile mass (kg), $S = \sigma h_2$, σ is breaking stress of backing plate (N/m²), h_1 is thickness of front plate (m), h_2 is thickness of back plate (m), ϵ is breaking strain of backing plate and

$$f(a) = \frac{M_p}{\left(M_p + (h_1\rho_1 + h_2\rho_2)\pi a^2\right)\pi a^2}$$

where $a = a_p + 2h_1$ (m) and ρ_1 and ρ_2 are densities of front and back plates respectively (kg/m^3).

Figure 12.18 shows the results of an investigation by Hetherington[8] into the optimisation of alumina-aluminium armours. Theoretical predictions based on the Florence model suggest that optimum performance, for a given areal density, will be achieved with a front plate to back plate thickness ratio (h_1/h_2) of approximately 2.5. Experimental findings support this prediction, but indicate that, provided $h_1/h_2 > 1$, the exact ratio is not critical.

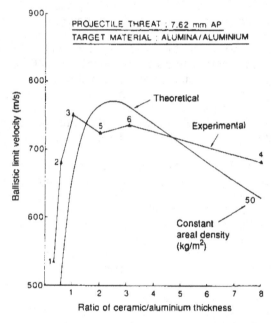

Figure 12.18 *Constant areal density optimisation of ceramic-faced armours (after Ref. 8)*

12.7 Numerical methods

The penetration prediction equations presented in Section 12.6 are useful for design purposes but do not provide a detailed understanding of the interaction between projectile and target. Computer-based methods can provide the researcher or designer with a great deal of valuable information about the penetration process. Such methods are particularly useful when dealing with rounds or targets having complex geometries or utilising novel materials.

Numerical methods are much more than a simple encoding of empirical equations of penetration. They appeal directly to the basic laws of physics – equilibrium, compatibility, conservation of mass, conservation of momentum – and draw on constitutive relationships for the materials involved. The majority of impact analysis codes are based on either the finite element method or the finite difference method of analysis. In the finite element method the penetrator and target are discretised into a large number of elements which meet at nodes. The degree of sophistication of a particular code depends largely on the nature of the elements employed. The displacement of a point at a particular location in an element is assumed to be determined by a polynomial function of its coordinates in that element. This polynomial function will contain coefficients (a, b, c . . .) the values of which are peculiar to that element. Strains may be deduced by differentiating the displacement function and stresses by invoking the constitutive relationships. The requirement that the conditions of equilibrium of forces and compatibility of displacements are met at every node leads to an array of simultaneous equations in which the only unknowns are the polynomial coefficients for each element. Solution for these coefficients leads to the evaluation of stresses, strains and displacements throughout the penetrator and target. By repeating this analysis during the incremental advance of the penetrator, a complete penetration history is developed.

By contrast, the finite difference method provides an approximate solution of a generalised governing equation for the body. Derivatives in the equations of continuum dynamics are replaced with difference approximations leading to a solution of stress, strain and displacement as a function of location and time.

The method by which a code keeps a record of a penetration process is often incorporated into the code designation. In a *Lagrangian* code, the progress of discrete elements of mass is monitored, whereas an *Eulerian* code monitors the flow of material into and out of cells within a fixed grid overlying the whole spatial domain.

When computer code output is supported by trials data it provides the design engineer with a powerful design tool. With development in both software and hardware, the computer code now offers a cost effective means of analysing terminal ballistic events. It is unlikely, however, that such a level of sophistication will be warranted in the context of structural protection. A fuller description of code types and capabilities is given by Zukas in his book *Impact Dynamics* listed in the bibliography at the end of this chapter.

12.8   Protective strategies

Provided the weight and bulk penalties can be tolerated, protection for static installations against ballistic threats can normally be provided by a sufficient thickness of concrete. Where rear-face spall constitutes a hazard, steel plating may be affixed to trap the debris. Where weight and bulk are critical considerations, metallic plate armours can provide an effective solution. Some further weight saving may be achieved by inclining armours due to the geometric and disruptive effects described in Section 12.5.4. Consider the requirement, described in Figure 12.19, to provide protection against a horizontal threat to a vertical vulnerable area of height h and breadth, perpendicular to the plane of the diagram, b. Trials have shown that a thickness t_n of armour is required to defeat the threat at normal and that the equivalent protection factor at obliquity θ is k_θ. The reduction in mass achieved by inclining the armour is

$$\rho b h \left(t_n - \frac{k_\theta t_n}{\cos \theta} \right)$$

Thus a weight *saving* only results if $k_\theta < \cos \theta$. This is equivalent to saying that the geometric effect offers no weight saving and that weight saving only comes from the disruptive effect. In practice the disruptive effect is only found in the subhydrodynamic regime and so inclining armour in this situation delivers no weight saving against hydrodynamic threats.

Explosive reactive armour (ERA) has proved an effective counter to shaped-charge rounds. An incoming jet detonates the explosive in a thin metal-explosive-metal sandwich. The jet is greatly degraded by the effects of all three layers of the sandwich rendering it ineffective against its intended target. Although a viable proposition for main battle tanks, ERA would only find application in structures under very specialised circumstances.

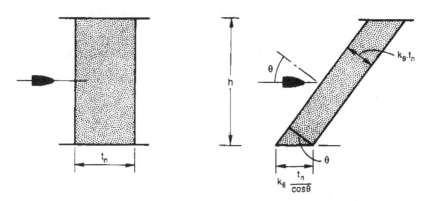

Figure 12.19 *Two alternative methods of protecting a vertical, vulnerable area*

Composite protective systems can offer enhanced protection when weight, bulk or cost are dominant considerations. Steel-concrete-steel and concrete-gravel-concrete sandwich systems have proved effective in structural applications. The incorporation of polymer or steel fibres into concrete has also proved useful in reducing spalling. Ceramic-faced armours offer some of the lightest protection currently available and will find increasing application for mobile targets.

12.9 References

[1] Recht R.F., *Ballistic Perforation Dynamics of Armor-Piercing Projectiles*, NWC TP4532, Naval Weapons Center, China Lake, California (1967)

[2] Stilp A.J. and Hohler V., *Long Rod Penetration Mechanics – High Velocity Impact Dynamics*, Chapter 5, Jonas A. Zukas (ed.). John Wiley & Sons, New York (1990)

[3] MOD, Assessment, Strengthening, Hardening, Repair and Demolition of Existing Structures. *Military Engineering* Vol. IX, Army Code No 71523 MOD, D.MOD SY, London (1992)

[4] Barr P., *Guidelines for the Design and Assessment of Concrete Structures Subjected to Impact*. Report of the CEGB, NNC and UKAEA Coordinating Committee for Structural Dynamics, Issue 5 (1987)

[5] Young C.W., Depth Prediction for Earth Penetrating Projectiles. *Journal of Soil Mechanics and Foundations Division*, ASCE Vol. 95, No. SM3 (1969)

[6] White M.P. (ed.), *Terminal Ballistics Part III. Effects of Impact and Explosion*. Summary Technical Report of Division 2, National Defense Research Committee, Washington, DC (1946)

[7] Florence A.L., *Interaction of Projectiles and Composite Armour Part II*. Stafford Research Institute, Menlo Park, California. AMMRG-CR-69-15 (1969)

[8] Hetherington J.G., The Optimisation of Two Component Composite Armours. *International Journal of Impact Engineering*. Vol. 12, No. 3 (1992)

12.10 Bibliography

Zukas J.A., Nicholas T., Swift H.F., Greszczuk L.B. and Curran D.R. *Impact Dynamics*. John Wiley & Sons, New York. (1982)..

Protective Construction Design Manual. Report No ESL-TR-87-57. Air Force Engineering and Services Center, Tyndall Air Force Base, Florida (1989).

Symbols

a	radius of base of conoid
a_p	radius of penetrator
A_x	presented area of penetrator at target surface
A, A_p	cross-sectional area of penetrator body
D	diameter of penetrator
e	thickness of slab to prevent perforation
E	young's modulus of target
f	coefficient of dynamic friction
f_{cu}	concrete cube strength
h_1	front plate thickness
h_2	back plate thickness
K	bulk modulus of target
L	length of penetrator
M	projectile mass
N	nose performance coefficient
n	ratio of round nose length to round diameter
s	thickness of slab to prevent scabbing
S	soil constant

t	plate or armour thickness
t_n	thickness required to defeat round at normal
t_θ	thickness required to defeat round at obliquity θ
u	velocity of base of crater
V	current velocity
V_o	impact velocity
V_{50}	ballistic limit velocity
v	rod speed
x	1 Distance from tip of projectile
	2 Penetration
α	half-cone angle of penetrator nose
ϵ	breaking strain of backing plate
θ	obliquity
ρ	density
ρ_p	density of penetrator
ρ_t	density of target
ρ_1	density of front plate
ρ_2	density of back plate
σ	breaking stress of backing plate
σ_y	yield stress of target material
τ	shear strength of target material
ψ	yaw

13 Buried structures

13.1 Introduction

Since life on Earth began, members of the animal kingdom have sought refuge underground. Recognising the potential of soil as a protective medium, man too has built underground structures to afford shelter from his enemy. Soil's quality of performance in this role derives from its inertia, its ability to diffuse load, its capacity to dissipate energy in plastic deformation and, of course, its obscuration of the target.

This chapter draws upon those elements of soil mechanics which are necessary to support an analysis of buried structures based on the theorems of *engineering plasticity*. This approach to design will enable the engineer to derive approximate solutions to a very wide range of problems quickly and with relative ease.

13.2 Essential soil mechanics

The integrity of a soil mass depends upon the soil's ability to withstand *shear stresses*. This ability, the *shear strength* of the soil, is determined by many factors including:

- the geological origin of soil particles
- the particle size distribution
- the fraction of the soil mass which is occupied by voids (i.e. air and water)
- the degree of saturation of the soil
- the loading history of the soil

A specific element of soil, in a specific location, under a specific set of *in situ* loads, will have a certain value of shear strength. Unfortunately, attempts to measure the shear strength of the soil, either *in situ* or by removing a sample of the soil to a laboratory, will result in disturbing the sample or changing the ambient state of stress. Nonetheless a range of laboratory tests exists which can be used to obtain shear strength parameters for use in analysis and design.

Soil is said to be a three phase medium, comprising solids, water and air. An imposed system of stress, which can be thought of as a combination of shear stress and isotropic stress, will be supported jointly by the soil particle matrix and the pore water. Water has zero shear strength and so any applied shear stress will be resisted exclusively by the soil matrix. The pore water pressure and the isotropic (i.e. *effective*) stress in the soil matrix together support the imposed isotropic stress.

The resistance of a soil mass to a system of applied shear stress depends, therefore, on the ability of the soil matrix to withstand shear stress, which in turn depends on the frictional interaction between the particles. The results depicted in Figure 13.1, obtained from a programme of triaxial tests on clay, show the characteristic dependence of shear strength τ upon effective stress σ'.

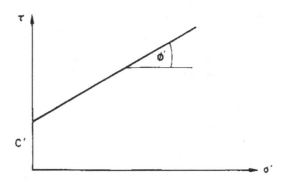

Figure 13.1 *Shear strength characteristics of soil*

The shear strength of the soil is said to have a cohesive element, represented by c', and a frictional element, represented by ϕ'. The equation of the graph

$$\tau = c' + \sigma' \tan \phi' \tag{13.1}$$

is known as the Mohr Coulomb failure criterion.

If an increment of load is suddenly applied to a saturated soil mass, the shear stress component resulting from the increment will have to be borne by the soil matrix. The isotropic stress will, however, be taken initially by the pore water and subsequently transferred to the soil matrix. The transfer of load is desirable since, as evidenced in Figure 13.1, this will strengthen the matrix. The *rate* of transfer is determined by the flow of water through the matrix and will therefore be very quick in sands and very slow in clay.

For this reason an increment of isotropic stress does not enhance the short-term shear strength of a saturated clay, which is therefore said to be *cohesive* in nature. In contrast, an increment of isotropic stress does enhance the short-term shear strength of sand and so it is said to be *frictional* in nature.

13.3 Concepts of plasticity

The Mohr Coulomb criterion, described in Section 13.2 above, provides a framework for describing the shear strength characteristics of soils. We adopt the criterion to describe the conditions which apply at the threshold between elastic and plastic regimes of behaviour, that is as a *yield criterion*.

The theory of plasticity was originally developed to account for post-yield behaviour of metals. Figure 13.2 shows a typical stress strain curve for mild steel obtained in a standard uniaxial tension test. A linear elastic region terminates at the yield point, followed by a region of strain hardening in the plastic regime. Superimposed on the curve is a bilinear relationship representing a *linearly elastic-perfectly plastic* model of behaviour. A *perfectly plastic* material is one which exhibits neither strain hardening nor strain softening. In other words, the yield criterion is satisfied throughout the plastic regime.

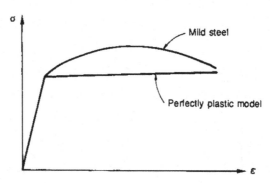

Figure 13.2 *Linearly elastic-perfectly plastic model of behaviour*

It is apparent from Figure 13.2 that, for the perfectly plastic material, we no longer find a one-to-one relationship between stress and strain in the plastic region. The strain which results from the application of loads to a perfectly plastic material is determined, not by the magnitude of the loads, but by the amount of work done by the loads.

13.4 Plasticity theory applied to soils

The stress state at a point in a soil mass is fully described by the nine stresses defined in Figure 13.3a. By notionally rotating the element, it is always possible to find an orientation of the element in which the shear stresses vanish to zero. The faces of the element are now *principal planes* and the direct stresses (σ_1, σ_2, σ_3) acting on the planes (Figure 13.3b) are the *principal stresses*. A convenient representation of the stress state is achieved by plotting a point with coordinates (σ_1, σ_2, σ_3) in principal stress space (Figure 13.4).

An element which is free of stress will be represented by a point at the origin and a point which represents the state of stress in an element, which is subjected to isotropic stress, will lie on the space diagonal OA along which $\sigma_1 = \sigma_2 = \sigma_3$. The point S clearly represents the state of stress in an element on which the three principal stresses are not equal. The further S is away from the line OA, the greater is the imbalance in the principal stresses and the greater will be the shear stresses, or the *deviatoric stress*, on the element. If,

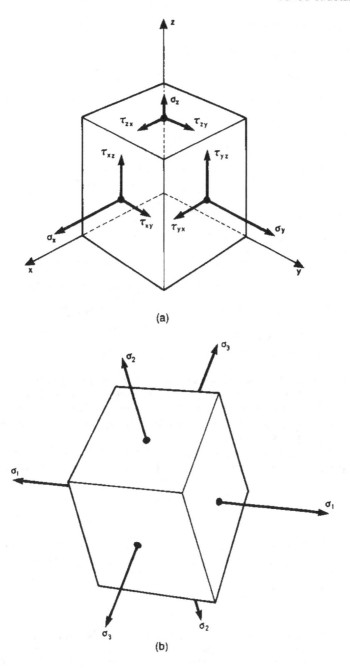

Figure 13.3 *(a) Generalised stress state in a soil mass; (b) principal planes and principal stresses*

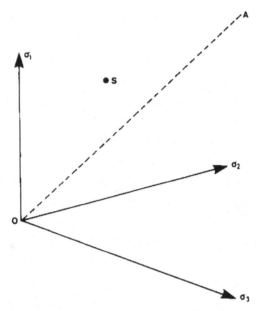

Figure 13.4 *Principal stress space*

by changing the applied stresses, we move S progressively further from the line OA, we will eventually bring the soil to yield. Indeed, if we were to conduct a large number of such experiments on the same soil, plotting in principal stress space all the points at which the soil yields, we could generate a surface as in Figure 13.5.

This is an example of a *yield surface* for it represents, for a particular soil, all states of stress which would bring the soil to yield. Having established a yield surface for a particular soil, we can depict test paths on the same diagram.

Example 13.1
A sample of saturated clay, whose yield surface is shown in Figure 13.5, is placed in a triaxial cell and, with drainage from the sample prevented, is subjected to a cell pressure of 100 kNm^{-2}. The ram is now driven down, shearing the sample to failure. Add a stress path to Figure 13.5 to depict the stresses imposed on the sample during this test.

Solution
The stress path comprises two straight lines (Figure 13.6). The point C has coordinates (100, 100, 100) and the line OC represents the imposition of the cell pressure. CY represents the increase of σ_2 while σ_2 and σ_3 are held constant at 100 kNm^{-2}. When the stress path reaches the yield surface at Y the soil yields. It is worth observing here that the test described above is a quick undrained triaxial test and so, for reasons described in Section 13.2, we would expect the shear strength of the soil to be unaffected by the isotropic stress level. Moreover, if the soil is isotropic, it should exhibit the same strength irrespective of the orientation of the principal planes of the applied stress

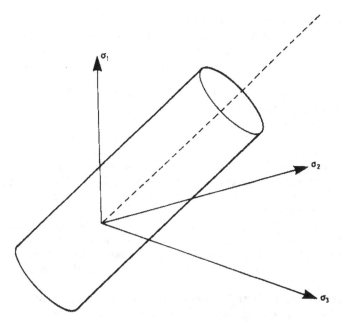

Figure 13.5 *Depiction of a yield surface in principal stress space*

field. These two considerations indicate that the yield surface for the quick undrained behaviour of a saturated clay should indeed be a cylinder as shown in Figure 13.6.

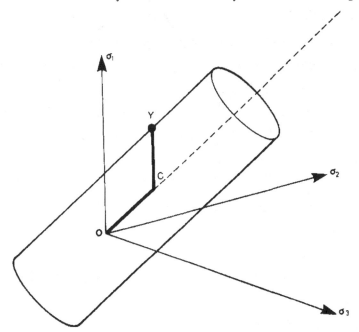

Figure 13.6 *Stress path for conventional triaxial test*

In practice many soils are frictional in nature. The key characteristic of a frictional soil is that its ability to support shear stresses (i.e. deviatoric stress) increases as the isotropic stress on the soil matrix increases. This means that the further the stress state is taken up the space diagonal, the greater will be the deviation from the space diagonal which can be tolerated before yield. This gives rise to the conical surface, known as the *Extended Von Mises* failure surface, shown in Figure 13.7.

It has already been argued (Section 13.3) that, for a perfectly plastic material, the magnitude of the plastic strain caused in a system is determined by the work done on the system by the external loads. Additionally, many materials closely follow a further model of behaviour known as the *normality rule*. The normality rule relates the magnitudes of the principal strain rates($\dot{\varepsilon}_1^P, \dot{\varepsilon}_2^P, \dot{\varepsilon}_3^P$) during plastic deformation to the yield surface in principal stress space as follows:

Firstly, axes of principal strain rate are superimposed on the axes of principal stress (Figure 13.7). A stress path is constructed to represent the loading which has brought the soil sample to yield (OCY). A normal is now constructed to the yield surface at Y, the point where the stress path strikes the surface.

This normal can be thought of as a *plastic strain rate vector*, whose direction indicates the relative magnitude of the principal strain rates.

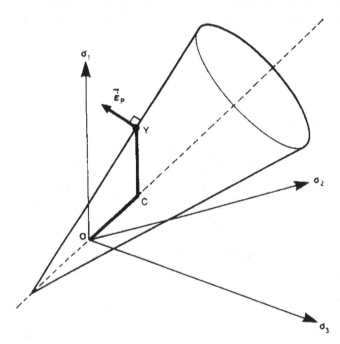

Figure 13.7 *Extended Von Mises failure surface*

A material where behaviour is modelled by the normality rule is said to have an *associated flow rule*. If the equation of the yield surface is written in the form $f(\sigma_1, \sigma_2, \sigma_3) = 0$, then the normality rule can be expressed in the following form:

$$\dot{\varepsilon}_1^P = \lambda \frac{\partial f}{\partial \sigma_1}$$

$$\dot{\varepsilon}_2^P = \lambda \frac{\partial f}{\partial \sigma_2}$$

$$\dot{\varepsilon}_3^P = \lambda \frac{\partial f}{\partial \sigma_3} \tag{13.2}$$

where λ is a scalar proportionality factor.

Example 13.2

The Coulomb failure criterion $\tau = c + \sigma \tan \phi$ can be written in the form

$$\sigma_1(1 - \sin \phi) - \sigma_3(1 + \sin \phi) - 2c \cos \phi = 0 \tag{13.3}$$

1 Prove that a Coulombic material must dilate during plastic deformation.
2 A 100 mm tall cylindrical sample of a saturated soil with $c = 100$ kNm^{-2} and $\phi = 20°$ is placed in a triaxial cell. With cell pressure constant and drainage permitted, the sample is brought to yield by driving the ram down at a constant rate of 2 mm per minute. Assuming the soil to be a perfectly plastic, Coulombic material, with an associated flow rule, determine the post yield diametral and volumetric strain rates.

Solution

It is important to notice, firstly, that compressive stresses are taken to be positive as are strains associated with reduction in size or volume.

1 From Equation 13.2,

$$\frac{\dot{\varepsilon}_1^P}{\dot{\varepsilon}_3^P} = \frac{\left[\dfrac{\partial f}{\partial \sigma_1}\right]}{\left[\dfrac{\partial f}{\partial \sigma_3}\right]}$$

From Equation 13.3 it follows that

$$\frac{\dot{\varepsilon}_1^P}{\dot{\varepsilon}_3^P} = -\left(\frac{1 - \sin \phi}{1 + \sin \phi}\right) \tag{13.4}$$

For the plane strain condition

$$\dot{\varepsilon}_2^P = 0$$

and

$$\dot{\varepsilon}_v^P = \dot{\varepsilon}_1^P + \dot{\varepsilon}_3^P$$

$$\therefore \quad \dot{\varepsilon}_v^P = \frac{-2 \sin \phi}{1 - \sin \phi} \cdot \dot{\varepsilon}_1^P \tag{13.5}$$

Alternatively, for the triaxial cell

$$\dot{\varepsilon}_2^P = \dot{\varepsilon}_3^P \quad \text{and} \quad \dot{\varepsilon}_v^P = \dot{\varepsilon}_1^P + 2\dot{\varepsilon}_3^P$$

$$\therefore \quad \dot{\varepsilon}_v^P = -\left(\frac{1+3\sin\phi}{1-\sin\phi}\right) \cdot \dot{\varepsilon}_1^P \tag{13.6}$$

It will be observed that, for all non-zero values of ϕ, the volumetric plastic strain rate is negative for both the plane strain and the triaxial cell condition. Thus for a Coulombic material, with an associated flow rule, plastic distortion will be accompanied by an increase in volume (i.e. dilation).

2 The sample, initially 100 mm tall, reduces in length by 2 mm per minute. Thus

$$\dot{\varepsilon}_1^P = 2\% \text{ per minute}$$

but from Equation 13.4 $\dot{\varepsilon}_3^P = -\dfrac{1+\sin 20^\circ}{1-\sin 20^\circ} \cdot 2\%$ per minute and therefore

$$\dot{\varepsilon}_3^P = -4.1\% \text{ per minute}$$

from Equation 13.6

$$\dot{\varepsilon}_v^P = -6.2\% \text{ per minute}$$

An important result is obtained by applying the normality rule for a Coulombic failure surface in σ, τ space (Figure 13.8). Following a parallel procedure to that adopted in Figure 13.7, axes of plastic direct strain rate $(\dot{\varepsilon})^P$ and plastic shear strain rate $(\dot{\gamma}^P)$ are superimposed on the axes of direct stress (σ) and shear stress (τ). It can be seen that shear strain will be accompanied by a negative direct strain (i.e. a dilation) and

$$\frac{\dot{\varepsilon}^P}{\dot{\gamma}} = -\tan\phi \tag{13.7}$$

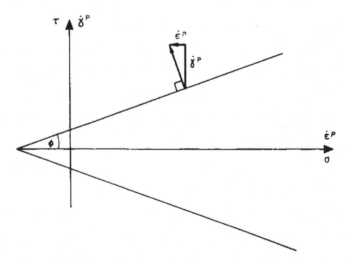

Figure 13.8 *Normality rule for Coulombic material*

Figure 13.9 shows a layer of plastically deforming Coulombic material, the upper surface of which has velocity components u and v and the lower surface of which is stationary.

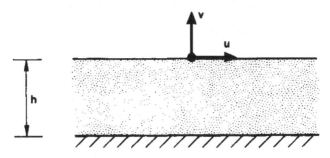

Figure 13.9 *Layer of plastically deforming Coulombic material*

By definition

$$\dot{\gamma}_p = \frac{u}{h} \quad \text{and} \quad \dot{\varepsilon}^P = -\frac{v}{h}$$

and substituting in Equation 13.7 gives

$$\frac{v}{u} = \tan\phi \qquad\qquad\qquad (13.8)$$

This result implies that the velocity vector of the upper surface of the layer makes an angle ϕ with the surface, where ϕ is the angle of friction of the soil. This finding will play an important part in our analysis of the stability of buried structures in frictional soils.

13.5 The limit analysis method of engineering plasticity

Engineering plasticity provides a method for analysing soil-structure systems, which have either complex geometry or complex loading or both. The method actually involves performing two quite separate analyses of the soil-structure systems which yield *upper and lower bound limits* on the collapse load for the system.

Consider, for example, the excavation depicted in Figure 13.10 adjacent to which is applied a surface pressure q. By applying both lower and upper bound methods of analysis to the problem, the true collapse load, q, of the system is bracketed between lower and upper bounds. The lower bound method will always produce an underestimate of the collapse pressure and therefore provide a safe or conservative basis for design. The upper bound method will always produce an overestimate of the collapse pressure and so constitutes an unsafe basis for design. By refining the bounds, it is possible to bracket the collapse load within a narrower range, thus improving the

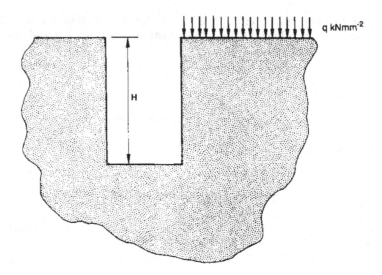

Figure 13.10   *Excavation in soil mass*

precision of the method. The engineer will cease expending effort in refin-ing the bounds when the discrepancy between upper and lower collapse load estimates is comparable with other inherent uncertainties (e.g. errors in determination of soil strength parameters).

The limit analysis method can also be used to determine limitations on geometry. Suppose, for example, that Figure 13.10 represents a scheme for an excavation in the vicinity of an existing structure. The foundation loads, represented by q, are already determined and the seminal question is 'how deep dare I make the excavation?' The upper bound estimate of H will, of course, be bigger than the lower bound estimate, the former being an unsafe and the latter a safe estimate of the permissible depth of excavation.

13.5.1   The lower bound method of analysis

The lower bound method is based on a 'postulated' or 'guessed' stress distribution within the soil mass. In reality, the actual stress distribution, which results from imposed loads and self-weight, may be very complex. The engineer's guess, on the other hand, will be much less complicated and based on zones or regions of constant or linearly varying stress. The postu-lated system must, however, obey the following two rules:

1   The stress system must be in equilibrium with the external loads.
2   The conditions of equilibrium must be satisfied internally throughout the system.

Figure 13.11 depicts an admissible stress system, judged by the above criteria, for the excavation problem of Figure 13.10. Since there is no support to the excavation the horizontal stress in zones 1 and 5 is set to zero. The vertical stress at a depth z in these zones is determined by the contribution

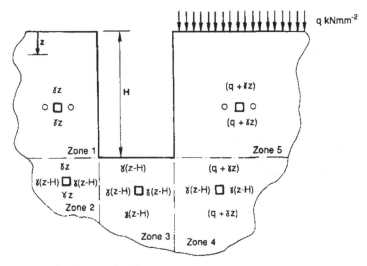

Figure 13.11 *Postulated stress distribution*

from external loads and self-weight. This is given by $\gamma_b z$, where γ_b is the bulk unit weight of the soil. Similarly, the vertical stress on an element at depth $z - H$ in zone 3 is given by the product of the bulk unit weight γ_b, and the depth of cover, $z - H$. The horizontal stress in zone 3 is not, however, determined by boundary conditions or self-weight and we arbitrarily decide to put zone 3 into isotropic compression. The condition of internal equilibrium with neighbouring zones then dictates the stress systems for zones 2 and 4.

The next stage is to examine the postulated stress system for potential failure. For simplicity we consider the short-term stability of the soil mass which we assume to comprise saturated soil with $\phi = 0$ and $c = c_u$. Zone 3 will not fail since it is in a state of isotropic compression and failure will occur in either zone 5 or zone 6 before zones 1 and 2 due to the imposed load. The worst site in zone 5 is at the bottom of the zone where $z = H$, the Mohr's circle for which is shown in Figure 13.12.

At any depth in zone 4, the diameter of the Mohr's circle is also $q + \gamma_b H$. Varying z simply slides the Mohr's circle from left to right without changing its size. Having obtained the Mohr's circles for the most critical sites in the system, we now notionally increase the vertical load until the Mohr's circles touch the yield surface (Figure 13.13).

At this stage, $q + \gamma_b H = 2c_u$ i.e.

$$q = 2c_u - \gamma_b H \tag{13.9}$$

This is a lower bound estimate on the value of q which will cause the system to collapse.

All lower bound estimates are, by definition, conservative (i.e. safe) estimates. This means that in reality a pressure greater than the value of q predicted by Equation 13.9 could be supported by the system. Just how

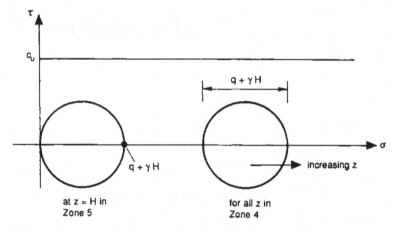

Figure 13.12 *Mohr's circle representation of postulated stress system*

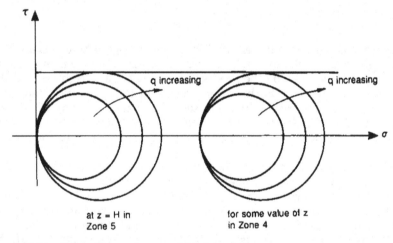

Figure 13.13 *Effect of increasing q on Mohr's circles*

conservative the estimate actually is can be gauged by comparing it with an upper bound estimate of the collapse load for the same system.

13.5.2 The upper bound method of analysis

The upper bound method is based on 'hypothetical' or 'assumed' mechanisms of collapse of the soil/structure system. In reality, collapse of the system will come about by the mechanism in which the least possible amount of energy is dissipated. Consequently, any mechanism proposed by an engineer as a method by which collapse could occur will be less energetically efficient than that adopted by nature. It is for this reason that the upper bound method always overestimates the amount of load which can be supported by the system. The engineer is, therefore, at liberty to select any geometrically feasible (i.e. *compatible*) collapse mechanism although, of course, the more nearly the hypothetical mechanism resembles reality, the better will be the estimate which results.

We then imagine the system collapsing at a slow, steady, constant speed and observe that the work done by the external loads acting on the system and gravity acting on the self-weight of the system must be dissipated in the soil and structure during the collapse. For convenience we consider the amount of work done and the amount of energy dissipated in the system *per unit time*. In other words we compare the power being put into the system with the power being dissipated in the system, leading to the following equation:

$$\begin{pmatrix} \text{Rate of work done} \\ \text{by external loads} \end{pmatrix} + \begin{pmatrix} \text{Rate of work done by} \\ \text{gravity on self-weight} \end{pmatrix} = \begin{pmatrix} \text{Power dissipated} \\ \text{in structure} \end{pmatrix} + \begin{pmatrix} \text{Power dissipated} \\ \text{in soil} \end{pmatrix} \quad (13.10)$$

Before we apply this method to the problem of the excavation discussed in Section 13.5.1, we need to identify candidate failure mechanisms for soil masses and establish methods for determining the power dissipated in them. The majority of failure mechanisms use rigid blocks of unyielded soil sliding over one another. The interfaces between them are thin layers of yielded soil which have suffered gross distortion. These are normally referred to as *surfaces of intense shear*. The adoption of this concept is supported by laboratory experiments and field observations of actual soil failures. Figure 13.14 shows a range of such mechanisms.

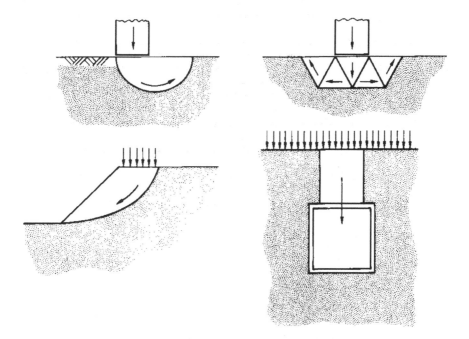

Figure 13.14 *Collapse mechanisms employing surfaces of intense shear*

Consider now a surface of intense shear in Coulombic material of thickness h, length l and unit depth (Figure 13.15). The surface of intense shear separates two rigid, unyielded blocks of soil, the lower stationary and the upper moving with velocities u and v parallel and perpendicular to the surface as shown. Equation 13.8 implies that the resultant velocity of the upper block makes an angle ϕ with the surface of intense shear. This observation has two important implications. Firstly, for frictionless ($\phi = 0$) materials (e.g. quick undrained behaviour of saturated clays) there is no separation of the blocks since there is no dilation of the material in the surface of intense shear. This means that both plane and cylindrical slip surfaces are admissible for this class of soils. Secondly, for soils with friction ($\phi \neq 0$) every point in the upper block moves off at an angle ϕ to the surface of intense shear. Only plane surfaces and logarithmic spirals permit this behaviour and are therefore the only admissible failure surfaces for this class of soils. Pursuing the general case for a (c, ϕ) soil, the rate of work done by the applied stresses $\dot{D}$ is given by

$$\dot{D} = \tau.l.1.u - \sigma.l.1.v$$

Substituting from Equation 13.8 leads to

$$\dot{D} = lu(\tau - \sigma \tan \phi) \tag{13.11}$$

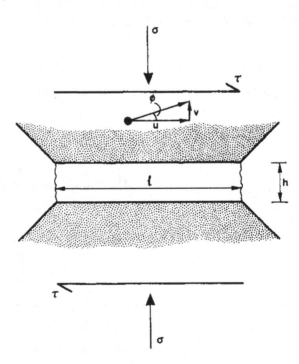

Figure 13.15 *Surface of intense shear in Coulombic material*

Substituting for τ from Equations 13.1 and 13.11 gives

$$\dot{D} = clu \tag{13.12}$$

Equation 13.12 gives the power dissipated in a surface of intense shear in any soil.

In analysing the collapse of buried structures, it is often convenient to employ a mechanism based not on surfaces of intense shear but on *regions of homogeneous shearing*. By a similar argument to that given above it can be shown that the rate of energy dissipation *per unit volume* is given by

$$\dot{d} = c\dot{\gamma} \tag{13.13}$$

where $\dot{\gamma}$ is the shear strain rate to which the region is subjected (Figure 13.16).

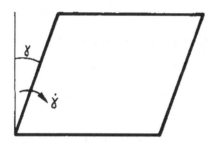

Figure 13.16 *Region of homogeneous shearing*

Example 13.3

For the excavation shown in Figure 13.10, derive an upper bound estimate for the pressure q to cause collapse.

Solution

The hypothetical collapse mechanism comprises a plane failure surface at an angle θ to the vertical (Figure 13.17). Since the soil is frictionless, the velocity of the sliding block, V, is parallel to the failure surface. Considering unit depth, the force acting on top of the sliding block is $q \cdot H \tan \theta$. The rate at which the external loads do work is equal to the product of this force and the component of the block's velocity in the direction of the force, i.e. $V \cos \theta$. Thus:

Rate of work done by external loads $= q \cdot H \tan \theta \cdot V \cos \theta$.

The mass of the sliding block equals $\frac{\gamma_b}{2} H^2 \tan \theta$ and the rate at which gravity does work on the system is $\frac{\gamma_b}{2} H^2 \tan \theta \cdot V \cos \theta$. The power dissipated in the surface of intense shear is (Equation 13.12).

$$\dot{D} = c_u \frac{H}{\cos \theta} \cdot V$$

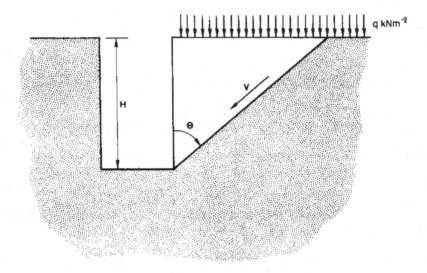

Figure 13.17   *Hypothetical collapse mechanism for excavation problem*

Applying Equation 13.10 and simplifying gives

$$q = \frac{2c_u}{\sin 2\theta} - \frac{\gamma_b H}{2} \tag{13.14}$$

We now vary θ to find a minimum value for q, the lowest upper bound available from this mechanism. This occurs at $\theta = 45°$ giving

$$q = 2c_u - \frac{\gamma_b H}{2} \tag{13.15}$$

The true collapse value of q will lie somewhere between the bounds indicated by Equations 13.9 and 13.15.

13.6   The limit analysis method applied to buried structures

In practice it is frequently more difficult to construct a lower bound solution than an upper bound solution for a buried structure. Indeed this is true for most problems and for this reason upper bound solutions are relatively common in civil engineering – yield line analysis of concrete slabs; plastic analysis of portal frames; slip circle analysis of earth embankments are all examples of the genre which are rarely recognised as such.

The concerning aspect of their prevalence is, of course, that they are, by definition, *unsafe*. If, however, engineering judgement, or experimental evidence, indicates that the selected mode of collapse is very close to that which would occur in practice, the upper bound estimate will be close to the true collapse load of the system. We start by developing an upper bound solution for a buried rectangular concrete structure.

Example 13.4

Figure 13.18 shows the cross-section of a long, reinforced concrete culvert, buried at a depth d in soil with bulk unit weight γ_b and quick undrained shear strength parameters c_u and ϕ_u. The system is simultaneously subjected to a line load of P kNm^{-1} and a pressure of q kNm^{-2}. Derive an expression for the fully plastic moment per metre run (M_p) of the culvert in order to provide a factor of safety F against collapse.

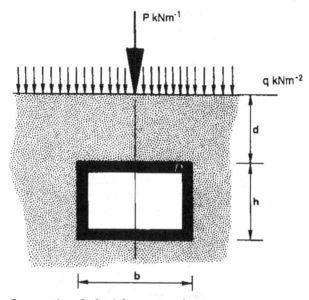

Figure 13.18 *Cross-section of a buried concrete culvert*

Solution

In searching for the correct solution to the problem we need to examine all of the possible mechanisms of collapse. Figure 13.19 shows some of the candidates. Each candidate mechanism should be the subject on an upper bound analysis but, as an example of the method, we will use our engineering judgement and select mechanism (a) for further study.

The roof slab of the culvert collapses through the formation of three yield lines or hinges running along the length of the slab (Figure 13.20). The rate of rotation of the slab is θ rad secs^{-1}. Invoking Equation 13.10 and considering 1 metre run of the culvert:

$$\text{Rate of work done by external loads} = P \cdot \frac{b}{2}\dot{\theta} + 2q \cdot \frac{b}{2} \cdot \frac{b}{4}\dot{\theta}$$

The second of these terms is obtained by doubling the product of the force on one-half of the roof $q \cdot \frac{b}{2}$ and the velocity with which the resultant force descends $\frac{b}{4} \cdot \dot{\theta}$. Similarly, the rate of work done by gravity on the soil is obtained by doubling the product of the weight of soil above one-half of the roof and the velocity with which its centre of gravity descends, i.e.

$$\text{Rate of work done by gravity on soil} = 2 \cdot \gamma_b \cdot d \cdot \frac{b}{2} \cdot \frac{b}{4}\dot{\theta}$$

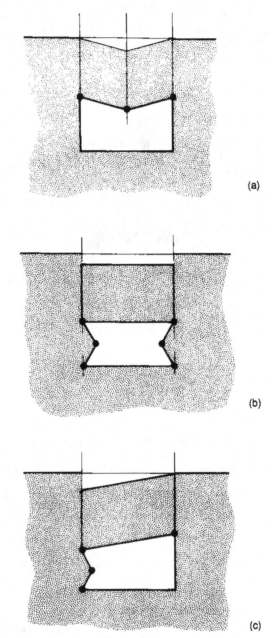

Figure 13.19 *Possible collapse mechanisms*

Clearly gravity also does work on the roof slab itself. If the mass per unit area of the slab is m kgm^{-2}, the contribution from this source becomes $2m\frac{b}{2} \cdot \frac{b}{4}\dot{\theta}$. This term is normally negligibly small when compared with the rate of work done in the soil mass and will therefore be omitted.

When a 1 m long hinge in a slab, with fully plastic moment per metre run of M_p kNm/m, is rotated at a rate of $\dot{\theta}$ radians per second, the power dissipated in the hinge is $M_p\dot{\theta}$ kW. The central hinge in Figure 13.20 rotates through 2θ when the edge hinges rotate through θ so the total power dissipated per metre in the structure is $4M_p\dot{\theta}$ kWm^{-1}.

The soil fails in two regions of homogeneous shear with a shear strain rate $\dot{\gamma} = \dot{\theta}$. From Equation 13.13, the power dissipated in shearing the soil is

$$2 \cdot c_u \cdot \dot{\theta} \cdot \frac{b}{2} \cdot d$$

Equating the terms gives

$$P \cdot \frac{b}{2} + q\frac{b^2}{4} + \gamma_b\frac{db^2}{4} = 4M_p + c_ubd \tag{13.16}$$

from which M_p can be found and factored by F to provide the required safety factor. Strictly, the correct method for including a factor of safety, however, recognises that the terms on the left of Equation 13.16 are driving the collapse whereas those on the right are resisting it. Thus

$$F = \frac{4(4M_p + c_ubd)}{2Pb + qb^2 + \gamma_bdb^2} \tag{13.17}$$

i.e. $$M_p = \frac{1}{4}\left\{\frac{F}{4}\left(2Pb + qb^2 + \gamma_bdb^2\right) - c_ubd\right\} \tag{13.18}$$

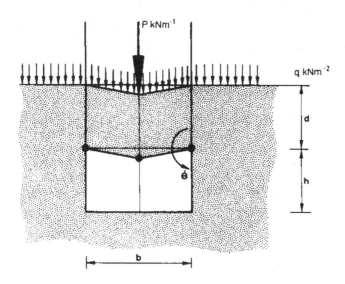

Figure 13.20 *Selected collapse mechanism*

The great virtue of the method of analysis employed in Example 13.4 is that it can be applied to a wide range of problems with a variety of loadings, geometrical configurations and soil types.

Moreover the form of Equation 13.17 enables us readily to determine the influence, on the integrity of a soil-structure system, of key factors including:

- depth of burial
- imposed loading
- dimensions of the structure
- strength of the structure
- strength of the soil

13.7   Blast and ballistic attack of buried structures

One of the primary reasons for burying a structure may be to afford protection from blast and ballistic attack. Whether or not a ballistic attack will damage a buried structure directly will depend on the ability of the round to penetrate soil and concrete. Methods for predicting the penetration of projectiles into such materials are discussed in Chapter 12.

Alternatively, a shell may be designed to penetrate the soil in the vicinity of a buried structure and then explode, causing damage to a structure through the transmitted groundshock. Methods for predicting the pressures delivered to the structure by this process are described in Chapter 7 whilst methods for analysing the transient response of the structure are detailed in Chapters 8 and 9.

Airborne blast waves, however, will result in a transient pressure applied to the surface of the ground. The effect of this transient pressure will always be less than the effect would be of the peak overpressure applied statically to the system. A convenient and safe method of design, therefore, is simply to use the peak overpressure in place of q, the imposed surface loading, in situations similar to that described in Figure 13.18.

13.8   Concluding remarks

This chapter has sought to equip the reader with a flexible and powerful method for analysing the structural integrity of any buried structure. In adopting this approach much detailed empirical knowledge on the response of buried structures has been omitted. This body of knowledge, which complements the treatment given here, is comprehensively covered by Bulson in his book 'Buried Structures' detailed in the bibliography.

13.9   Bibliography

Bulson P.S. *Buried Structures – Static and Dynamic Strength*. Chapman and Hall, London. ISBN 0-412-21560-8 (1985)

Calladine C.R. *Engineering Plasticity*. Pergamon Press, Oxford. ISBN 0-08-013-969-8 (1969)

Wai-fah Chen, *Limit Analysis and Soil Plasticity*. Elsevier Scientific Publishing Company, Amsterdam. ISBN 0-444-41249-2 (1975)

Symbols

c'	effective stress shear strength parameter – cohesion
c_u	undrained shear strength parameter – cohesion
$\dot{D}$	rate of work done
$\dot{d}$	rate of energy dissipation per unit volume
d	depth of burial
F	factor of safety
h	thickness of shear layer (Fig. 13.15)
l	length of shear layer (Fig. 13.15)
M_p	fully plastic moment per metre run
m	mass per unit area of roof slab
P	line load
q	surface pressure
u	horizontal velocity component (Fig. 13.9)
v	vertical velocity component (Fig. 13.9)
z	depth coordinate
γ_b	bulk unit weight of soil
$\dot{\gamma}^P$	plastic shear strain rate
$\dot{\varepsilon}^P$	plastic direct strain rate
$\dot{\varepsilon}_1^P \dot{\varepsilon}_2^P, \dot{\varepsilon}_3^P$	principal plastic strain rates
σ	direct stress
σ'	effective stress
$\sigma_1, \sigma_2, \sigma_3$	principal stresses
τ	shear stress, shear strength
ϕ	angle of friction
ϕ'	effective stress parameter – angle of friction
ϕ_u	angle of friction obtained in undrained test

14 Protective design

14.1 An overview of the approach to design

This chapter provides an outline guide to design methodologies for the prevention of damage to structures due to dynamic loads such as those caused by explosives or impacting projectiles. A number of measures may be necessary depending on the nature of the applied loading, the form of the structure under consideration and the properties of materials used in construction.

Prior to loading, measures can be taken to prevent or minimise damage to the structure. These can be described as providing either protective *strengthening* or *hardening*. The aim of both of these measures is to increase the resistance of the structure to the applied loading. However, they deal with different effects, which may be caused by different types of loading.

Strengthening can be defined as those measures taken to increase the overall strength and stability of a structure under the specified loading. The objective of strengthening is the prevention of a major structural failure, rather than the reduction of damage to individual elements. An example of strengthening is the creation of a high level of redundancy in a structure, so that failure of one or two members will not lead to collapse.

Hardening is concerned with increasing the resistance of a structure, or elements within it, to damage from impacting projectiles. It is thus concerned with minimising damage to an element or part of the structure. This could be achieved by treatment of the structure itself, or perhaps by the provision of an additional protective barrier.

Before the above procedures can be carried out for an existing structure it is essential that the state of the structure be assessed. The purpose of the assessment procedure will depend on whether the structure is damaged or undamaged. In the case of undamaged buildings, it is necessary to assess the initial strength and hardness of the structure in order to decide what enhancements are required in order to resist the anticipated blast or impact loading. For damaged structures, it is necessary to assess the extent of the damage before selecting either the appropriate repair method or, if damage is excessive, choosing to demolish the building.

14.2 General philosophy of protection

The first step before taking any protective measures is to consider the threat to which the structure might be subjected. This is often the hardest aspect of the entire process. For example, the structure may come under 'general attack' from either aerial bombardment or ground-based attack from artillery or mortar fire. 'Global' effects from such threats could include the generation of damaging blast overpressures or substantial groundshock loads. Alternatively the structure may be subject to specific localised effects. For instance, the structure may need to withstand ballistic penetration by a range of weapons or be required to resist explosive penetration from a charge detonated in contact with the structure. Consideration may need to be given to penetration by shaped charges and the threat from multiple hits by fragments is often significant.

In all of these aspects, consideration must be given to the response not only of the structure but also of equipment and personnel located within the structure and to assess what level of response is acceptable. This will then allow consideration of the level of protection required. Questions such as whether the structure damage should be limited to a minor level (indicating a high degree of protection/strengthening) or whether a higher level of damage can be tolerated, in which case protective measures need not be so comprehensive. Linked with structure response is the response of personnel within the structure and consideration must be given to what is considered an 'acceptable' level of injury or fatality.

14.3 Guidelines for new protective structures

Consider now the structure itself and its capacity for blast resistance. In Table 14.1 the performance of various structural forms and construction materials is given in a rough ranking order. Best performance – low risk of both local and progressive failure – features at the top of the list with poor performance at the bottom.

In all situations it is, of course, desirable to investigate the area likely to be exposed to any blast and minimise it. It is often worth considering the provision of some form of external protection. For example, it may be desirable to use a burster slab over an underground structure, though it could be that the costs of such a measure may be prohibitive. The use of blast walls adjacent to above ground structures is commonplace but there are questions about the efficiency of such structures. It is vital that such walls should be not only robust but that their location relative both to the position of the threat and the buildings that they are protecting is given due consideration. If the blast wall is too far from the structure the blast wave from the attack will re-form behind the wall and could produce a significant load on the structure. If the attack is located at too great a distance from the wall, there will be little energy absorption by the wall itself through wall deformation.

It may be important to consider effects of groundshock waves produced by a device exploding either on the ground or after penetrating into the

Table 14.1

| Form | Risk of failure | | Comments |
	Local	Progressive	
Underground/low profile arch/mass construction	Low	Low	Excellent airblast resistance
RC box with cross-walls	Low	Low	Corridors around periphery, cladding, protected air intakes (for personnel protection)
RC frame	Varies	Low	Depends on cladding, floor construction and joint details
Steel frame	Varies	Low	
Prestressed concrete frame	Varies	Varies	Joint details very important
Masonry/brickwork	High	High	Use cross-walls to increase rigidity
Industrial (large panel)	High	High	Poor blast resistance

ground adjacent to a structure. The provision of a shock attenuating medium between the source and the structure, often referred to as 'back-packing', can go some way to providing structural protection.

It may be worth incorporating a sacrificial component in the structure such as a sacrificial roof. If a device detonates on top of the sacrificial component energy is absorbed by the 'dummy element' which also has the effect of maintaining some stand-off to the 'real' structure beneath. The measures noted above are illustrated in Figure 14.1.

If no external protection can be provided then it is desirable to design key elements to resist blast pressures with the overall structure designed to avoid progressive collapse in the event of their removal. Structural elements should be tied together to resist likely blast pressures. Thus, continuity of reinforcement and the provision of so-called 'blast links' in reinforced concrete structures is important for the absorption of blast energy.

Consider now the ballistic resistance of a structure. Of course, it is vital to arrest missile penetration by providing adequate wall and floor thickness in the chosen material of construction which might be in the form of steel plates or reinforced concrete slabs. It would also be prudent to provide measures to prevent scabbing of the structural material remote from the impact point of any projectile to increase personnel protection. This can be achieved in reinforced concrete construction by the use of steel anti-scabbing plates firmly attached to the inside of the structure wall. The use of steel or polypropylene fibre reinforcement in the concrete can also prove effective in reducing the risk due to scabbing.

The approach to the formulation of a successful new blast resistant structure has been summarised in Table 14.2, adapted from Baker *et al.*[1], which lists 12 check points and questions which the designer must address. Note particularly item 4 – the structure must be able to carry all 'normal'

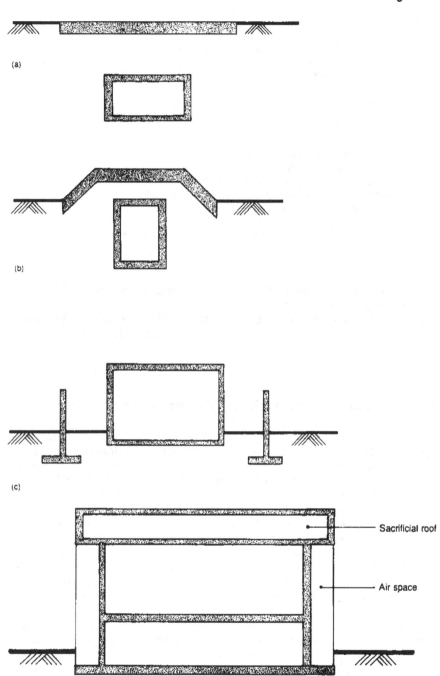

(a)

(b)

(c)

(d)

Sacrificial roof

Air space

Figure 14.1   *Possible methods of structural protection*

Table 14.2

	Blast-resistant design sequencing
1	Where is the structure to be sited?
2	What layout is required?
3	What is structure function?
4	Design for static loads
5	Assess threat
6	Calculate blast loads
7	Assess any fragment characteristics
8	Calculate any fragment impact loads
9	Assess level of any crater ejecta loading
10	Assess whether groundshock may be significant
11	Carry out preliminary sizing
12	Analyse response as simply as possible

loading and conventional static design criteria should be employed to ensure integrity of the structure.

Reference 1 presents flow diagrams which outline the paths to the design of structures to resist externally and internally generated blast loading in considerable detail.

14.4 Guidelines for protecting existing structures

The foregoing relates to the design of new structures which offer protection against blast and associated loadings. To upgrade existing structures the approach requires some modification. Firstly, it is necessary to categorise the structure under consideration to enable the engineer in the field to rate rapidly the ability of the structures to resist the various dynamic loadings associated with a particular threat. Categories could include underground arches (for example, cellars and tunnels), low rise buildings in precast concrete or masonry, reinforced concrete or steel building frames, towers or masts and water retaining structures. Secondly, it will be necessary to assess the level of damage likely from a particular threat. If they are available, this could be done by the use of damage curves in the form of pressure-impulse diagrams. It will then be necessary to provide the required protection against the threat.

To protect against blast overpressures the use of blast walls has already been mentioned. Reinforced concrete cast *in situ* could be deployed but consideration could be given to the use of precast concrete planks or concrete blockwork built between rolled steel joists used as columns cast into the ground. It may also be possible to make use of suitably buttressed blockwork. Sacrificial roofs have been advocated as integral parts of new structures. Here the use of scaffolding for temporary protective purposes or the addition of trussed timber rafters for permanent protection could be a viable option. Another option to provide extra stand-off could be to encase a vulnerable part of the structure in precast cladding.

To counter ballistic impact and other forms of impulsive load, key structural elements can be protected by the use of steel skins tied to the parent structure with the inclusion of an intermediate energy absorbing layer. For example, a sand infill will absorb the jet from a shaped-charge attack. The use of stiffening elements can increase resistance where the strength of beam/column joints can be increased by the use of steel plates bolted or adhesively bonded to the structure.

Against ballistic penetration the use of arresting nets and screens can prove effective. Interlocking or mortar-bonded concrete block walls and rapidly assembled precast concrete panel walls could be considered. It is worth noting, however, that plain concrete can produce fragments on the front face and a scab or spall which is blown off the rear face, thereby creating secondary missiles. The use of steel or polypropylene fibre reinforced concrete is advocated which is effective both in improving tensile strength and in energy absorption. The more conventional anti-scabbing measures such as the steel-concrete sandwich construction noted above are also applicable in the upgrade of existing structures.

14.5   The overall cost of protective strengthening

In designing and building a new protective structure consideration must be given to constraints on time and cost of construction as well as any constructional and technical restraints. Figure 14.2, adapted from Reference 2, shows the site layout for a hardened structure for which the threat has been assessed as a vehicle bomb.

A number of features of the outline design are worthy of note. Firstly, the structure is located with its outer walls at a minimum distance from the perimeter of the site. This minimum distance will be determined in the light of the perceived threat so that, should the structure be attacked, damage to the structure will be within the acceptable limits. The bollards are designed to prevent any breakthrough of the perimeter protective wall by a vehicle driven at speed. The vehicle barrier at the only entrance to the site is designed to have the same effect. Vehicular access is very restricted with the majority of visitors' vehicles being directed to the off-site parking.

The net result of the application of these measures will be an enhanced level of protection of a structure against a range of specified threats. It is instructive to compare the cost of providing protection with the cost of any loss that would be incurred. As will be seen from Figure 14.3, the cost of protection rises as the level of protection required increases with a very high cost being associated with total protection. As protection level rises so the cost of any loss will decrease such that, if the structure is 100% protected, no losses will be incurred. The sum of the two curves indicates that there will be an optimum level of protection corresponding to the minimum of the total cost.

This approach can be particularised somewhat if the problem of providing stand-off is considered. In the context of blast loading, the provision of

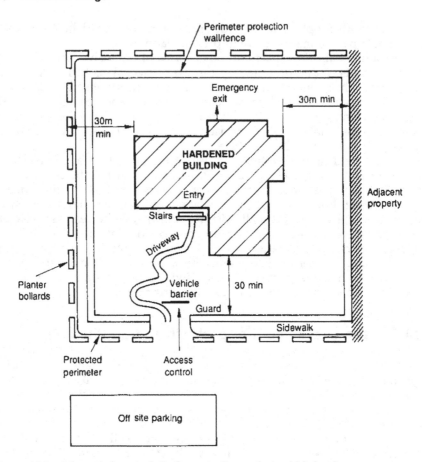

Figure 14.2 *Schematic layout of site for protection against vehicle bombs*

reasonable stand-off between the threat and the structure will lead to less structural damage. Figure 14.4 shows how, for a given charge weight W, as stand-off distance is increased so the cost of hardening diminishes in proportion to $(1/R^3)$, while the cost of the land necessary to provide stand-off increases at a rate proportional to R^2, where R is stand-off distance. The minimum total cost obtained by summation of the two curves gives the optimal hardness range though, if there are more funds to buy more land, protection can be increased by the provision of further stand-off to the point where, if stand-off is large enough, no hardening costs will be incurred because at this range the structure is inherently hard to the necessary degree.

Although the cost of strengthening individual elements of a structure may appear quite high when expressed as a percentage of the cost of that element alone, when expressed as a percentage of the total building cost it is considerably less. Figure 14.5 shows how increasing the thickness of a reinforced concrete wall by a factor of 3 may cost 50% more in terms of wall cost alone but only a 3% increase in cost for the total building. Similarly, a two-fold

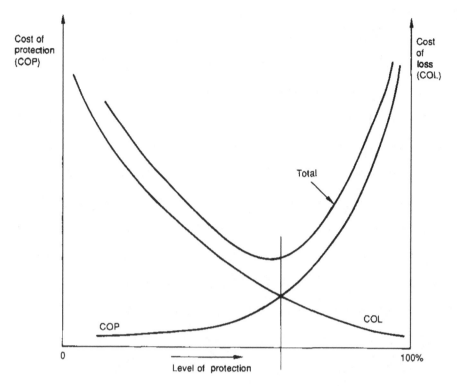

Figure 14.3 *Cost of protection vs protection level*

increase in the cost of glazing may only represent between a 5 and 10% increase when expressed in terms of the entire building.

Overall, substantial protection may be afforded by an increase in overall costs of the order of 5 to 10%. This is likely to represent the optimum position when balancing the cost of protection against the potential cost of the subsequent loss. A breakdown of costs associated with hardening is presented in Figure 14.6.

The lesson to be learnt from such analysis is that in reality a compromise between perimeter-to-structure distance and the level of protection designed into the structure combined with non-structural security measures would afford an acceptable level of protection with only a relatively minor cost increase compared with costs of a totally unprotected structure.

14.6 Outline of blast resistant design

14.6.1 Reinforced concrete compared with steel

Before illustrating the principles of design against blast loading it is important to highlight the differences between the response of reinforced concrete and that of steelwork. These are summarised in Table 14.3.

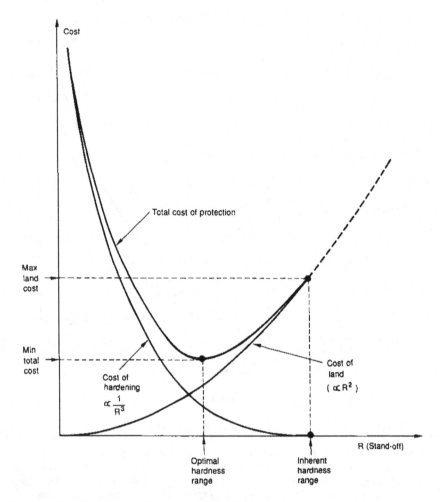

Figure 14.4   *Cost of protection vs stand-off distance*

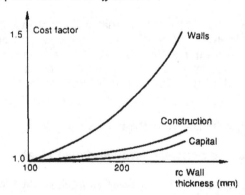

Figure 14.5   *Cost factors for increasing protective wall thickness*

Table 14.3

Reinforced concrete	Structural steelwork
Suitable for extremely close-in explosions resulting in massive structures.	Best suited to structures subjected to low pressure loadings (protection may be required).
Because of the generally 'stocky' construction the ultimate carrying capacity is fairly predictable.	Because of the generally slender sections, local instabilities and buckling can give unpredictable ultimate capacities.
Internal damping by massive cracking means little rebound under large deflections.	Up to 100% rebound under the lower pressure/longer duration loads means that strength must be supplied for full reversals of load.
Separate reinforcement for bending, for shear and for torsion gives a more clearly defined response.	The need for combined stress analysis can lead to a difficult problem of assessing stress levels under dynamic loads.
Sections generally change in a regular manner to facilitate formwork.	Stress concentrations at welds, notches and joints must be carefully considered so that full strength in the section can be generated.

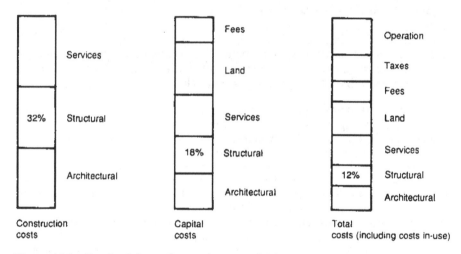

Figure 14.6 *Cost breakdown of protective strengthening*

The principle of design is essentially the same for both materials. The structure or structural element is characterised by an idealised elastic-plastic resistance function and, if the loading is from a blast wave from a condensed high explosive source, it is idealised as a triangular pulse of zero rise time. The element to be designed should be converted, wherever possible, to the appropriate single degree of freedom lumped-mass system.

14.6.2 *Blast resistant design in reinforced concrete*

The resistance function for a simple element such as a one-way spanning under-reinforced concrete section is usually represented as a graph of ultimate resistance per unit area r_u against rotation of the section at supports as shown in Figure 14.7 below.

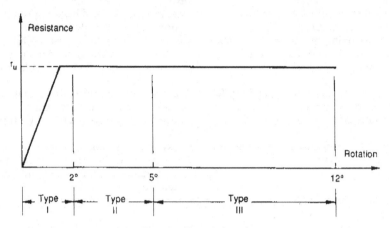

Figure 14.7 *Resistance-rotation function for reinforced concrete*

Before undertaking a detailed design it must be decided what constitutes acceptable performance for the element. This means identifying the level of damage that the element may sustain and is classified as Type I, II or III behaviour as defined in Reference 3 given below.

Type I: minor cracking $\quad 0 < \theta < 2°$
Type II: crushing and cracking $\quad 2° < \theta < 5°$
Type III: disengagement of concrete
$\qquad$ from reinforcement (spalling) $\quad 5° < \theta < 12°$

Using the ideas of Chapters 8 and 9, it is also necessary to establish whether the loading of the structure is likely to be in the impulsive, quasi-static or dynamic regimes. If the duration of the loading pulse t_d is a lot less than the response time of the structure T, the loading should be considered impulsive. This is generally the case for reinforced concrete structures loaded by blast waves from conventional munitions. This may be quantified by stating that loading is impulsive if either or both of the following criteria are met:

$$\frac{t_d}{T} < 0.1 \tag{14.1}$$

(which corresponds to the criterion of Equation 9.10) or

$$\frac{F}{r_u} > 11 \tag{14.2}$$

where F is the peak load produced by the blast.

The ultimate resistance of the section varies with the failure type chosen by the designer, section thicknesses and amount of reinforcement and the way the plastic or ultimate bending moment is calculated. An alternative way of expressing damage levels is to equate Types I, II and III with values of ductility ratio μ of about unity, 3 and 30 respectively, where ductility ratio is the ratio of the maximum deflection of the structure to the maximum elastic deflection. Linear deflections are sometimes easier to visualise than joint rotations in describing damage.

The principles described above are best illustrated by reference to a specific element. Consider the design of a protective blast wall of height H and width b (Figure 14.8) to be constructed around the periphery of a structure.

The blast wall is to be allowed to suffer considerable damage and in so doing absorb a lot of energy. Thus, a Type III element will be designed in which the rotation (at the wall base) may lie between 5° and 12°. Because the elastic response of the structure comprises only a small part of the overall response the resistance function of Figure 14.7 may be simplified to the resistance-deflection graph of Figure 14.9 in which X_m is the maximum (transient) deflection of the element.

The wall is subjected to a pressure-time loading which may be simplified to the linear decay shown in Figure 14.10, where p_r is the normally reflected overpressure and t_d is the duration of the positive phase.

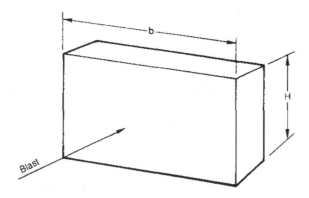

Figure 14.8 *Blast-loaded cantilever wall*

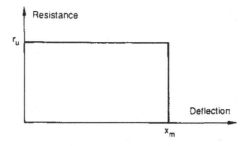

Figure 14.9 *Idealised resistance-deflection function for reinforced concrete*

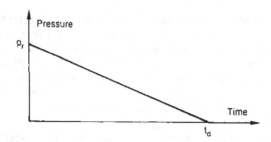

Figure 14.10 *Idealised blast load pressure-time history*

For impulsive loading t_m, the time for the structure to undergo its maximum response should be at least $3t_d$. Assuming that this is so, the reflected overpressure impulse i_r is required given by the area under the positive phase of the pressure-time curve equal to $p_r t_d/2$. The total impulse delivered to the wall will be $i_r Hb$. If the wall is of density ρ the mass of the wall will be ρHbd_c, where d_c is the effective depth of the wall which is also here taken as the overall wall thickness. The kinetic energy KE delivered to the wall is given by

$$KE = \frac{(i_r bH)^2}{2\rho bd_c H} = \frac{i_r^2 bH}{2\rho d_c} \tag{14.2}$$

The work done by the wall in deforming WD will be given by

$$WD = r_u bHX_m \tag{14.3}$$

If KE and WD are equated the term bH on each side of the equation cancels to leave

$$\frac{i_r^2}{2\rho d_c} = r_u X_m \tag{14.3}$$

where the group ρd_c could be described as the areal density of the wall. At this stage a further conversion is desirable by the inclusion of the load-mass factor K_{LM} to give

$$\frac{i_r^2}{2K_{LM}\rho d_c} = r_u X_m \tag{14.4}$$

Clearly, the larger the value attainable for X_m the larger the blast impulse that can be resisted by the element. The maximum deflection attainable depends on span, damage level and the amount and distribution of the reinforcement.

To ensure ductile behaviour in elements subject to impulse loading, two changes in reinforcement layout when compared with 'conventional' design must be made. Firstly, the element should be reinforced symmetrically which enables the compression reinforcement to carry all the compressive stresses once concrete in the compression zone has crushed and spalled.

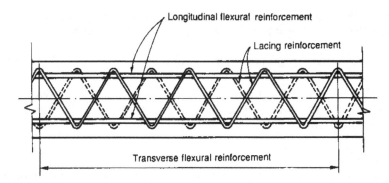

Figure 14.11 *Typical laced reinforcement arrangement*

Secondly, the main flexural steel and enclosed concrete should be 'laced' together using an arrangement such as that shown in Figure 14.11.

This method allows the strain hardening region of the steel stress/strain behaviour to be fully developed and also mobilises the shear strength of tensile steel and core concrete. It also has the effect of restraining the compression reinforcement from buckling and helps to spread out any effects of non-uniform loading.

Writing the area of the tensile steel as A_s and setting this equal to the area of the compression steel A_s', the ultimate moment of resistance of the wall M_u is given by

$$M_u = f_{ds}A_s d_c \tag{14.5}$$

where f_{ds} is the dynamic yield strength of the steel. Equation 14.5 can be rewritten in the form

$$M_u = \frac{A_s}{bd_c}f_{ds}bd_c{}^2 = \rho_v f_{ds}d_c{}^2 b \tag{14.6}$$

where ρ_v is the vertical reinforcement ratio. The ultimate moment can also be written in terms of r_u as

$$M_u = (r_u bH)\frac{H}{2} \tag{14.7}$$

if it is assumed that the wall deflects by rotation about the base and that the resistance is assumed to act at the centroid of the loaded area (here at $H/2$ from the base of the wall). Hence, by combining Equations 14.6 and 14.7 we obtain

$$r_u = \frac{2}{H^2}(\rho_v f_{ds}d_c^2) \tag{14.8}$$

The maximum deflection at the top of the wall X_m is given by

$$X_m = H\tan\theta \tag{14.9}$$

where θ is the rotation at the base of the wall. If this and the expression for r_u are substituted into Equation 14.4 we obtain the final equation

$$\frac{i_r^2}{2K_{LM}\rho d_c} = \frac{2}{H}(\rho_v f_{ds} d_c^2)\tan\theta \tag{14.10}$$

from which, if the loading is known, the steel requirements and overall wall thickness can be determined. Example 14.1 below illustrates the approach for a specific situation.

Example 14.1

A cantilever wall of height 3.6 m is loaded by a blast from a vehicle bomb comprising approximately the equivalent of 100 kg of TNT at a range of 3.8 m. The wall is re-inforced with steel of static yield strength 460 N/mm² and the vertical reinforcement ratio ρ_v is to be 0.5%. The overall density of the wall may be taken as 2400 kg/m³ and the dynamic increase factor for the steel is 1.2 (see below). The load-mass factor K_{LM}, may be taken as 0.66.

What steel will be required if the structure is to exhibit Type III behaviour?

The vehicle bomb may be idealised as a hemispherical charge of effective mass 1.8 times the actual mass, i.e. 180 kg. At a range of 3.8 m this represents a scaled distance Z of $3.8/180^{1/3}$ which is approximately 0.67 m/kg$^{1/3}$. From the scaled distance graphs of Figure 3.17 the reflected overpressure impulse i_r can be obtained as approximately 5100 Pa-s. The dynamic yield strength of the steel f_{ds} is taken as 550 N/mm² ($\approx 1.2 \times 460$) or 550×10^6 N/m². Substituting these values into Equation 14.10 gives

$$\frac{5100^2}{2 \times 0.66 \times 2400 \times d_c} = \frac{2 \times 0.005 \times 550 \times 10^6 \times d_c^2 \times 0.213}{3.6} \tag{14.11}$$

Therefore, on rearrangement

$$d_c^3 = \frac{5100^2 \times 3.6}{2 \times 0.66 \times 2400 \times 2 \times 0.005 \times 550 \times 10^6 \times 0.213} \tag{14.12}$$

from which $d_c = 293$ mm. The overall thickness of the wall T_c can then be taken as 425 mm to allow for cover and lacing. The area of steel per metre width of wall needed will be

$$A_s = 0.005 \times 293 \times 1000 = 1465 \text{ mm}^2/\text{m}$$

This can be achieved using 20 mm diameter bars at 200 mm centres on each face of the wall.

Similar calculations can be performed for Types I and II behaviour. The resulting wall will be more robust because of the requirement for reduced damage. Table 14.4 below shows the results of such calculations carried out assuming that the loading is still impulsive in form.

Table 14.4

Failure type	I	II	III
d_c (mm)	535	395	293
T_o (mm)	660	520	425
Reinforcement	25mm @ 200	25 mm @ 250	20 mm @ 200
X_m (mm)	125	320	765
Rotation	2°	5°	12°

14.6.3 Blast resistant design in structural steel

In the example above for concrete, the dynamic strength of the reinforcing steel was made greater than the static yield strength by multiplying by a dynamic increase factor which can be up to 1.9 depending on the rate at which loading is applied. Table 14.5 shows how the increase factor varies for structural mild steel and the resulting dynamic yield strengths in tension and shear.

Table 14.5

Time to reach yield stress	Ultimate dynamic yield stress f_{ds} (N/mm^2)	Dynamic shear stress f_{dv} (N/mm^2)
>1 second	f_y (250 N/mm^2)	f_v (155 N/mm^2)
100 ms – 1 sec	$1f_y – 1.1f_y$	155 – 172
10 ms – 100 ms	$1.1f_y – 1.6f_y$	172 – 250
1 ms – 10 ms	$1.9f_y – 1.6f_y$	250 – 295

Figure 14.12 shows similar information in graphical form expressed as a function of strain rate for two different grades of structural steel.

It would be unusual to design a protective structure from steel but, of course, steel structures are very likely to suffer what may be termed 'incidental' blast loads in an attack even though the steel structure may not be the main target. Possibly a better way of regarding what follows would be as an assessment guide rather than a set of design criteria for steel structures.

If a steel structure is to have any chance of survival it should only be called on to resist relatively low pressure blast loads of relatively long duration. Steel structures should, as in the case of concrete structures, be idealised as SDOF systems. Steel structures are categorised as either reusable (implying that they should only be required to sustain light damage) or non-reusable, where the structure sustains severe damage yet still retains its structural integrity. The maximum ductility ratios associated with these categories should be defined such that incipient failure is approached and significant deformations well into the strain-hardening range are allowed for good energy absorption. Thus, for reusable structures, μ_{max} is approximately 3 while for non-reusable structures μ_{max} is taken as about 6.

The design method outline is firstly to determine the blast loading history for the structure, the geometry of the element or member to be designed, the appropriate material dynamic properties and the ductility ratio range for which the structure is to be designed. An estimate of the maximum element resistance value R_u is then made. For reusable structures R_u is set to 1.0 F_1, while for non-reusable structures R_u is $0.8F_1$, where F_1 is the maximum value of blast loading applied which is assumed to be a triangular load-time history.

The next step is to determine the required value of the plastic moment of resistance M_p using the section modulus is the governing criterion. Thus for the two categories we have

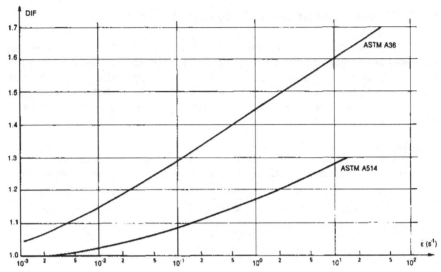

Figure 14.12 *Dynamic increase factor for yield strength of steel vs strain rate*

$$M_p = f_{dy}(s + z)/2 \quad \text{for} \quad \mu \le 3$$

$$M_p = f_{dy}z \quad \text{for} \quad 3 < \mu \le 6 \tag{14.13}$$

where s is the elastic section modulus and z is the plastic section modulus. Since even a reusable structure will undergo plastic deformation, the average of s and z is used in determining M_p. It should be noted that, of course, normal 'static' design criteria should also be checked at each stage. The member mass can now be found and converted to an equivalent mass by multiplying by K_{LM} the load-mass factor from Table 10.2. It is now possible to calculate T_N, the effective natural period of vibration of the element.

Calculation of R_u and F_1 can proceed, from which the ratio of load pulse duration to natural period (t_d/T_N) can be obtained from Figure 14.13 which shows how ductility ratio varies with this time ratio so that the ductility ratio supplied can be read for the calculated R_u/F_1 ratio.

Now the value of μ thus obtained can be compared with that previously chosen and if it is unsatisfactory the process can be repeated as necessary. Finally, a check should be made on the performance in shear of the element. Using the empirical result that I-sections achieve their fully plastic moment capacity provided the average shear stress over the full web area is less than the yield stress in shear, then

$$V_p = f_{dv} \cdot A_w \tag{14.14}$$

where V_p is the shear capacity of the section and A_w is the web area. The actual value of maximum shear force V must be less than V_p. Shear values may be determined from the dynamic reaction values for beam elements from Table 10.2. To illustrate this approach consider a specific example.

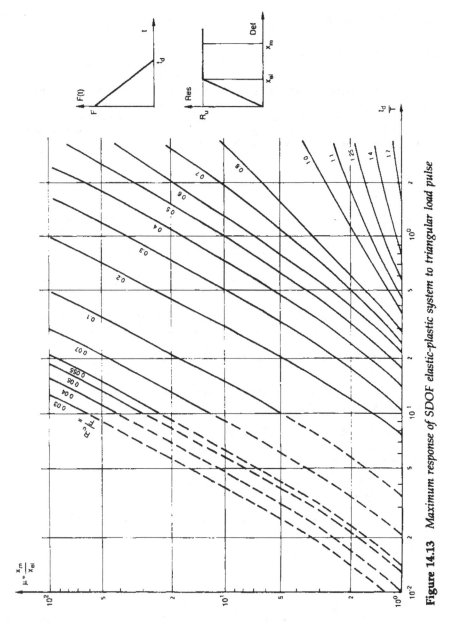

Figure 14.13 *Maximum response of SDOF elastic-plastic system to triangular load pulse*

Example 14.2

A simply supported steel beam of span 5.2 m is to be designed as a reusable structure to resist a particular blast loading. The beam may be thought of as one of a number at 1.4 m centres forming the roof of a structure which is clad with material producing a load (excluding the self-weight of the beam) of 23.4 kg/m². What will be the dimensions of the beam if the blast loading, idealised as a triangular load-time pulse, is as shown below in Figure 14.14? The dynamic increase factor for the steel may be taken as 1.3 and the static yield strength as 250 N/mm².

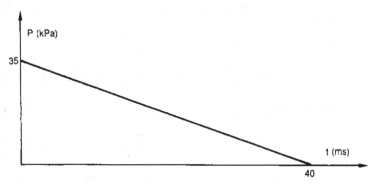

Figure 14.14   *Blast overpressure-time history for Example 14.2*

Since the structure is to be reusable the dynamic load ratio (the ratio of R_u to F_1) is 1.0. Hence the value of R_u is calculated as

$$R_u = 1.0 \times 35 \times 5.2 \times 1.4 = 254.8\,\text{kN} \tag{14.15}$$

Hence, the plastic moment of resistance M_p is given by

$$M_p = \frac{R_u L}{8} = \frac{254.8 \times 5.2}{8} = 165.6 \ \text{kN-m} \tag{14.16}$$

From Equation 14.13 for M_p, rearrangement will give, for a reusable structure, a value for $(s + z)$

$$s + z = \frac{2M_p}{f_{dy}} = \frac{2 \times 165.6 \times 10^3}{1.3 \times 250 \times 10^6} = 1.02 \times 10^{-3}\text{m}^3 \tag{14.17}$$

which is 1020 cm³. From tables of section properties of universal beams such as Reference 4, a 356 × 127 × 39 UB is found to meet the requirement with an $(s + z)$ value of $(653.5 + 571.8)$ equal to 1255 cm³ which exceeds the required 1020 cm³. Assuming that normal static design requirements are satisfied, the actual mass m carried by the beam can now be found as the sum of the self-weight of the beam and the weight of the cladding as

$$m = 39 \times 5.2 + 23.4 \times (5.2 \times 1.4) = 373.1\,\text{kg} \tag{14.18}$$

From Table 10.2 the elastic and plastic load-mass factors are 0.78 and 0.66 respectively, their average being 0.72. Thus, the period of oscillation of the equivalent system T_N is given by

$$T_N = 2\pi \sqrt{\frac{K_{LM} m}{K_e}} \qquad (14.19)$$

where K_e is the equivalent stiffness given by

$$K_e = \frac{384EI}{5L^3} \qquad (14.20)$$

where EI is the flexural rigidity of the beam of span L. Here, on substitution

$$K_e = \frac{384 \times 210 \times 10^9 \times 10\,087 \times 10^{-8}}{5 \times 5.2^3} = 1.157 \times 10^7 \text{N/m} \qquad (14.21)$$

Hence

$$T_N = 2\pi \sqrt{\frac{0.72 \times 373.15}{1.157 \times 10^7}} = 3.03 \times 10^{-2}\text{s} = 30.3\,\text{ms} \qquad (14.22)$$

Then, from Figure 14.13, since t_d/T_N is 1.32 (= 40/30.3), R_u is evaluated as

$$R_u = \frac{8M_p}{5.2} = \frac{8 \times 1225 \times 10^{-6} \times 250 \times 10^6 \times 1.3}{5.2 \times 2} = 3.062 \times 10^5 \text{N} \qquad (14.23)$$

and so R_u/F_1 is 1.2 and the ductility ratio μ is found to be 1.5. Since this is less than 3 the design is acceptable for a reusable structure.

Finally, a check on the shear performance should be made given that the dynamic yield stress in shear is f_{dv} given by

$$f_{dv} = 0.5 \times 1.3 \times 250 = 162.5\text{N/mm}^2 \qquad (14.24)$$

where the factor 0.5 is used to relate yield stress in tension to shear yield stress. Thus the ultimate shear capacity is

$$V_p = f_{dv} \cdot A_w = 162.5 \times (6.5 \times 352.8) = 3.726 \times 10^5 \text{N} = 372.6\text{kN} \qquad (14.25)$$

where A_w is the area of the web obtained from Reference 4. The maximum support shear V (the dynamic reaction generated from Table 10.2 or Equation 10.56) is given by

$$\begin{aligned} V &= 0.393R_u + 0.107F_1 \\ &= 0.393 \times 3.062 \times 10^5 + 0.107 \times 2.548 \times 10^5 \qquad (14.26) \\ &= 1.476 \times 10^5 \text{N} = 147.6 \text{ kN} \end{aligned}$$

which is less than the maximum allowed value of 372.6 kN and the design is acceptable.

14.7 Construction of blast resistant structures

Blast resistant structures often occur as parts of buildings and are commonly constructed by those who are traditionally concerned with appearance and finish. It should be clearly emphasised that it is structural considerations that are of prime importance in blast resistant buildings. Also it should be understood that the standards of workmanship required in blast resistant structures may differ somewhat from those which might be related to normal concrete construction. There should be an understanding by all parties concerned that the occasional defects in design or workmanship should be mitigated or corrected.

In construction of a blast resistant structure it is important to consider the following aspects of the process. Firstly, reinforcement splices should be designed so as not to coincide with any stress concentration regions. Secondly, the erection sequence of the reinforcement should be carefully planned to enable all steel to be correctly located. Thirdly, when pouring concrete it is important to plan the location of vertical and horizontal construction joints to avoid high stress concentration regions. Finally, it is important to plan the sequence of concrete pours to optimise the integrity of the structure. It should also be remarked that since a relatively high density of reinforcing steel may be required then, for structures such as walls, temporary staying of the reinforcement could be required to maintain adequate cover.

As mentioned above, construction joints should be carefully considered and should be avoided whenever possible as they are potentially weak sites. All unavoidable joints should be in low stress regions. If possible, vertical joints should be in areas of horizontal lacing and horizontal joints in areas of vertical lacing. Base level joints are generally used with horizontal joints in a wall and are only allowed if the wall height does not permit a full height pour. Finally, walls which intersect should be poured simultaneously.

It should also be noted that there may be access problems for vibration by virtue of the high density of lacing steel employed. Care must therefore be taken to ensure that adequate vibration has been carried out before access and monitoring becomes impossible. If, following stripping, it is found that defective work is present, breaking out and making good for isolated patches should be limited to what might be acceptable for work exposed to the sea. For example, edges should be defined by diamond sawing, sharp internal corners should be avoided and the patch should be of sufficient size to be fully engaged with the reinforcement. Surface making good should be prohibited. As a general guide the actual concreting process should be that which would be adopted for water-retaining structures. Because of the relatively large amounts of reinforcement, expansion joints are generally not required but may be incorporated in long buildings or those subject to extreme temperature changes. When pouring concrete, to maintain a minimum pour rate, multiple pouring crews may be needed.

14.8 Design codes

Although there is a British Standard BS 5628: The Structural Use of Masonry[5] which includes provision for accidental damage as summarised in Reference 6, the principal codes for structural design to resist the effects of blast loading are from American sources.

The document TM3-1300[3] has already been mentioned above and in Chapters 3 and 8. It should be regarded as the principal source for the design of protective structures. When used in conjunction with the document TM5-855-1[7] and the recent publication *Protective Construction Design Manual*[8] the designer has a comprehensive 'toolkit'. For the design of structures to resist the effects of nuclear blast loading the document TM5-856-1 through 9[9] is available, though a more readily available source containing the same approach is the *ASCE Manual of Engineering Practice No. 42.*[10]

14.9 References.

[1] Baker W.E., Cox P.A., Westine P.S., Kulesz J.J., Strehlow R.A., *Explosion Hazards and Evaluation* Elsevier. Amsterdam (1983)
[2] Hinman E.E., Physical Countermeasures for the Protection of Buildings against 'Car-bomb' Attack. Proceedings of Symposium on Securing Installations Against Car-bomb Attack. Arlington, Va (1986)
[3] US Department of the Army Technical Manual TM5-1300. Design of Structures to Resist the Effects of Accidental Explosions (1991)
[4] Howatson A.M., Lund P.G., Todd J.D., *Engineering Tables and Data.* (1st edition) Chapman and Hall, London (1972)
[5] BS 5628: Structural Use of Masonry: Part 1: Unreinforced Masonry (1978)
[6] Morton J., Accidental Damage, Robustness and Stability. Brick Development Association CI/SfB Fg2 (1985)
[7] US Department of the Army Technical Manual TM5-855-1. Fundamentals of Protective Design for Conventional Weapons (1987)
[8] Drake J.L., Twisdale L.A., Frank R.A., Dass W.C., Rochefort M.A., Walker R.E., Britt J.R., Murphy C.E., Slawson T.R., Sues R.H. *Protective Construction Design Manual.* Air Force Engineering Services Centre, Engineering and Services Laboratory, Tindall Air Force Base. ESL-TR-87-57 (1989)
[9] US Army Technical Manual TM5-856-1 through 9. Design of Structures to Resist the Effects of Nuclear Weapons (1965)
[10] American Society of Civil Engineers. Design of Buildings to Resist Nuclear Weapon Effects. *ASCE Manual of Engineering Practice No. 42* (1985)

Symbols

A_s	area of tensile steel
A'_s	area of compression steel
A_w	area of web
b	wall width
d_c	effective depth of section
EI	flexural rigidity of beam
f_{dv}	dynamic shear yield stress
f_{dy}	dynamic yield stress
f_v	static shear yield stress
f_y	static yield stress
F	peak blast load
F_1	peak blast load
H	blast wall height
i_r	reflected specific impulse
K_e	equivalent stiffness
K_{LM}	load-mass factor
KE	kinetic energy

L	span of beam
m	mass of material carried by beam as dead load
M_p	plastic moment of resistance
M_u	ultimate moment of resistance
p_r	peak reflected overpressure
r_u	ultimate resistance per unit area
R	stand-off distance
R_u	ultimate moment of resistance
s	elastic section modulus
t_d	duration of idealised triangular blast load
t_m	time taken to undergo maximum response
T	natural response period of structure
T_c	overall wall thickness
T_N	effective natural period of structure
V	shear force
V_p	shear capacity of beam
W	charge weight
WD	work done by blast load
X_m	maximum element deflection
z	plastic section modulus
θ	structure rotation
μ	ductility ratio
μ_{max}	maximum ductility ratio
ρ	density of structure material
ρ_v	vertical reinforcement ratio

15 Worked examples

15.1 Introduction

This chapter presents a series of worked examples to illustrate the principles introduced in this book. They are arranged in categories, for example, 'Blast Loading', 'Structural Response to Blast Loading', 'Buried Structures', etc. There could, however, be some overlap between sections in questions that require ideas from more than one chapter to be applied for complete solution.

15.2 Explosives

Example 1
An explosive charge of mass W is to be detonated on the surface of a frictional soil (sand) of friction angle θ and density ρ. The charge has a specific energy Q kJ/kg and a density Δ. Derive an expression for the volume of crater V that such a charge might produce. How might such an event be modelled in the laboratory?

Solution
Writing the most general form of relationship between the variables gives:

$$V = f[g, Q, \rho, W, \Delta, \phi] \tag{15.1}$$

This equation must be dimensionally correct so, when each quantity is expressed in terms of its fundamental or primary units of mass (M), length (L) and time (T), each side of the equation has the same primary units. The method due to Rayleigh leads to the formulation of a series of dimensionally correct terms which can then be expressed in terms of non-dimensional numbers (sometimes called Pi-groups) as demonstrated below. Writing the equation in dimensional terms:

$$[L^3] = [LT^{-2}]^\alpha [L^2 T^{-2}]^\beta [ML^{-3}]^\gamma [M]^\delta [ML^{-3}]^\epsilon [0] \tag{15.2}$$

Equations are now written for the indices of each of M, L and T:

$$M: \quad 0 = \gamma + \delta + \epsilon$$
$$L: \quad 3 = \alpha + 2\beta - 3\gamma - 3\epsilon$$
$$T: \quad 0 = -2\alpha - 2\beta \tag{15.3}$$

These equations have five unknowns and a decision must be made as to how to solve, in order that g and Δ appear only once in any set of Pi-groups. Thus we have

$$\beta = -\alpha$$
$$\gamma = -1 - (\alpha/3) - \epsilon$$
$$\delta = 1 + (\alpha/3) \tag{15.4}$$

Hence the original equation can be recast as

$$V = g^\alpha Q^{-\alpha} \rho^{-1-\alpha/3-\epsilon} W^{1+\alpha/3} \Delta^\epsilon \phi^0 \tag{15.5}$$

Collecting terms with the same index gives

$$V = \left[\frac{g}{Q}\left(\frac{W}{\rho}\right)^{1/3}\right]^\alpha \left[\frac{\Delta}{\rho}\right]^\epsilon \frac{W}{\rho} \tag{15.6}$$

which can be rewritten as a general equation for crater volume involving three Pi-groups:

$$\frac{V\rho}{W} = f\left[\frac{g}{Q}\left(\frac{W}{\rho}\right)^{1/3}, \frac{\Delta}{\rho}\right] \tag{15.7}$$

To gain more specific information it would now be necessary to conduct experiments. The appearance of gravitational acceleration in the second of the Pi-groups could present severe problems in meaningful small-scale modelling of a full-scale event unless a centrifuge were employed.

As an example consider the modelling of a 10 tonne surface burst of a condensed high explosive. If 10 grammes of the same explosive were to be used in a centrifuge experiment employing the same soil as in the real event, the requirement that the gravity Pi-group be the same in both model and full size event means that the centrifuge should produce a 100 g gravitational field. For a centrifuge with an arm of radius 5 m, this field can be realised if the speed of rotation is just over 2 revolutions per second.

15.3 Blast waves and blast loading

Example 1
What will be the peak static overpressure P_s, the positive phase duration T_s, the arrival time t_a and the specific impulse produced by a spherical charge of 25 kg RDX at a range of 5 m?

Solution
From Table 3.2, the TNT equivalence factor for RDX is 1.185 giving the TNT equivalent charge mass here as $1.185 \times 25 = 29.63$ kg. The scaled distance at a range of 5 m is $Z = 5/(29.63)^{1/3} = 1.62$. The required quantities can now be derived by entering Figure 3.8: $p_s = 2.6 \times 10^5$ Pa; $i_s = 340$ Pa-s; $T_s = 5.25$ ms and $t_a = 3.40$ ms.

Example 2
A 10 kg spherical charge comprises 7 kg of PETN and 3 kg of Torpex. The wall of a structure at 4 m from the charge is loaded face on by the blast wave produced by central initiation of the charge. What is the reflection coefficient for this loading?

Solution
Using the information of Table 3.2, 7 kg of PETN contains $7 \times 5800 = 40\,600$ kJ while 3 kg of Torpex contains $3 \times 7540 = 22\,620$ kJ making a total charge energy of 63 220 kJ which equates to $63\,220/4520 = 13.98$ kg TNT. At 4 m range from the wall the scaled distance $Z = 4/(13.98)^{1/3} = 1.66$. From Figure 3.8, p_s is read off as 2.4×10^5 Pa and, from Figure 3.17 p_r is found as 1.1×10^6 Pa. The reflection coefficient C_R is thus $p_r/p_s = 4.58$.

Example 3
What is the maximum range at which a peak static overpressure of 1 bar (15 psi) will be produced by a 10 kt weapon? What should the height of burst be to achieve this?

Solution
From Figure 3.22 for a 1 kt weapon, the maximum range at which 1 bar overpressure can be developed is 355 m if the weapon is initiated at a height of burst of 200 m. To scale these distances for a 10 kt weapon Equation 3.30 is used giving a scaling factor of $(10\,\text{kt}/1\text{kt})^{1/3} = 2.154$. Thus 1 bar overpressure will be developed at a range of $355 \times 2.154 = 765$ m if the height of burst is $200 \times 2.154 = 431$ m.

Example 4
A communications centre occupies an area which can be approximated to a 2 km diameter circle with the main operations buildings situated at one point on the perimeter with other equipment dispersed around the rest of the circumference. The main building will suffer complete collapse when it is loaded by a long duration overpressure pulse of 100 kPa (= 15 psi) while all other equipment is hardened to only 70 kPa (= 10 psi). Calculate the minimum yield of nuclear device necessary to achieve these pressures together with the location of its ground zero.

Solution
The furthest separation of the two targets is when both are at opposite ends of a diameter of the airfield. The device must be initiated at some height of burst along the diameter. If the distance from the main building is a metres and the distance from the other equipment is b metres, then

$$a + b = 2000 \tag{15.8}$$

Also, irrespective of weapon size the ratio $a : b$ must remain constant. Thus, from Figure 3.22 for a 1 kt device, the maximum range to produce 100 kPa is 355 m at a height of burst of 200 m. At this height of burst the range at which 70 kPa is generated is 430 m. Therefore

$$\frac{a}{b} = \frac{355}{430} = 0.826 \tag{15.9}$$

Solving Equations 15.8 and 15.9 gives $a = 904.5$ m and $b = 1095.5$ m. Since these results are for a 1 kt device, scaling up involves using Equation 3.30. Thus, the necessary yield is

$$W = 1 \times \left(\frac{1095.5}{430}\right)^3 = 16.5\,\text{kt} \tag{15.10}$$

Example 5
A nuclear explosion may be regarded as an instantaneous release of a given amount of energy E concentrated at a point. The radius of the blast wave so produced, r, is a function of this energy, the density ρ and the pressure p of air at ambient conditions and the time elapsed after detonation t. Show that a general relationship between these parameters exists of the form

$$r\left(\frac{\rho}{Et^2}\right)^{1/5} = f\left(p\left(\frac{t^6}{E^2\rho^3}\right)^{1/5}\right) \tag{15.11}$$

In the near to medium field the influence of atmosphere may be ignored. Write down the relationship between the remaining parameters.

A nuclear device releases 90 000 GJ of energy in air of density 1.225 kg/m³. After 2 seconds the radius of the blast wave is 745 m. How long will it take a blast wave produced by a weapon ten times as big to achieve a radius of 1 km if atmospheric conditions are the same? Comment on your findings.

Solution
Using the approach detailed in Section 15.2, Example 1, the general relationship between the important parameters

$$r = f(E, \rho, p, t) \tag{15.12}$$

can be recast in the non-dimensional form of the question by solving the M, L and T equations in terms of the index chosen for pressure (so that p appears in only one of the two Pi-groups).

If the effect of ambient pressure conditions may be ignored, only the first Pi-group remains which must be a constant, say, A. Thus

$$r\left(\frac{\rho}{Et^2}\right)^{1/5} = A \tag{15.13}$$

Substituting $E = 9 \times 10^{13}$J, $\rho = 1.225$ kg/m³, $t = 2$ s and $r = 745$ m gives $A = 0.952$. Then, for the larger weapon, substitution into Equation 15.13 gives, on rearrangement, that the time for the wave to achieve a radius of 1 km is 1.32 s. This indicates that, even though the distance travelled by the wave is longer, the larger weapon produces a faster travelling wave and the time to reach 1 km is shorter than for the smaller weapon wave to reach 745 m.

Example 6
Calculate the arrival time of the blast wavefront, t_a, 5 m away from a spherical charge of high explosive detonated in air where the speed of sound a_o is 350 m/s if measurements of peak side-on overpressure p_s were made using an array of seven gauges as detailed below.

R(m)	2.5	3.0	3.5	4.0	5.0	6.0	8.0
p_s (bar)	6.15	3.85	3.57	1.87	1.10	0.72	0.42

Solution
From the data above it is possible to calculate the Mach number of the wavefront $M(= U_s/a_o)$ using Equation 3.19. These values are 2.47, 2.07, 1.99, 1.60, 1.38, 1.26 and

1.16 corresponding to the gauge locations. If the wavefront speed U_s $(= Ma_0)$ is written as the rate of increase in distance from the charge dR/dt, then, on rearrangement

$$\int_0^{t_a} dt = \frac{1}{a_0} \int_0^5 \frac{dR}{M} \tag{15.14}$$

This equation enables arrival time at 5 m to be evaluated by, for example, graphical integration. If $1/M$ is plotted against R as shown in Figure 15.1 the area beneath the curve from $R = 0\,\text{m}$ to $R = 5\,\text{m}$ when divided by a_0 will give the required arrival time.

The area is found to be about 2.35 (in metres) and so arrival time is $(2.35/350) \times 1000 = 6.71$ milliseconds.

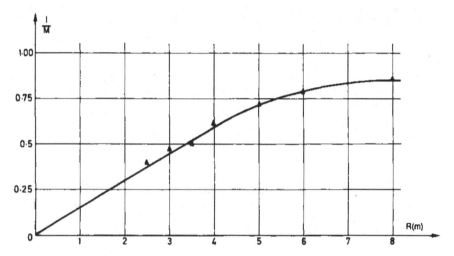

Figure 15.1 *Graphical solution to Example 6*

Example 7

A chamber for handling explosives is a cube of internal dimension 4 m. Half of the roof of the structure is made of frangible material to allow venting in the case of any accidental detonation. Evaluate the vent time if detonation of a charge of 5 kg of TNT occurs in the chamber. What will be the gas pressure impulse that the structure will need to resist if the chamber is to survive? Atmospheric pressure may be taken as 10^5 Pa and the speed of sound in air as 350 m/s.

Solution

The chamber has a volume $V = 4 \times 4 \times 4 = 64$ m^3 so the charge/volume ratio $W/V = 5/64 = 0.078$. From Figure 4.14, the peak quasi-static pressure P_{QS} is read off as 3.3×10^5 Pa. The scaled peak quasi-static pressure

$$\bar{P}_1 = \frac{P_{QS} + P_0}{P_0} = 4.3 \tag{15.15}$$

Then, from the figure, $\bar{\tau}$ is obtained as 0.7. Given that the vent area is $2 \times 4 = 8\,m^2$ and the area of walls and roof is $5 \times (4 \times 4) = 80m^2$, the parameter α_e is 0.1 and, with a total chamber surface area of $96\,m^2$, vent time can be evaluated as 13.33 milliseconds. Finally $\bar{i}_g$ is read as 1.0 from which gas pressure impulse i_g is obtained as 1904 Pa-s.

15.4   Underwater Explosions

Example 1
If the specific energies of HMX and TNT are 5680 and 4520 kJ/kg respectively and given that for HMX the index α in Equation 5.11 for peak shock pressure ($P_m = KZ'^{\alpha}$ where $Z' = R/W^{1/3}$) is 1.16, calculate the peak pressure at 15 m from a 125 kg HMX charge. K for TNT is 52.5 (to give pressure in megaPascals).

Solution
The K value for HMX can be evaluated by using the specific energy ratio of HMX and TNT. Thus $K_{HMX} = 52.5 \times (5680/4520) = 65.85$. Therefore the peak pressure generated at a range of 15 m is

$$p_m = 65.65 \times \left[\frac{15}{125^{1/3}}\right] = 18.4\,MPa \tag{15.16}$$

Example 2
Calculate the pressure in an underwater shock wave 1 ms after arrival if measurement is being made at 15 m from a 125 kg charge of TNT and the decay behind the shock front is given by the equation

$$P(t) = P_m \exp{(-t/\theta)} \tag{15.17}$$

where θ is the time for the pressure to fall to P_m/e and is given by the empirical equation

$$\theta = K_\theta W^{1/3} Z^{0.22} \tag{15.18}$$

where K_θ for TNT is 92 to give θ in microseconds.

Solution
Using Equation 5.11 and the data of Table 5.1 for TNT gives the peak pressure at 15 m from the 125 kg TNT charge as 15.17 MPa. Substitution into Equation 15.18 gives the parameter $\theta = 586$ microseconds. Thus, using Equation 15.17 the pressure 1 ms after shock arrival is

$$p(1ms) = 15.17 \times \exp\left[-\frac{1000}{586}\right] = 2.75\,MPa \tag{15.19}$$

Example 3
What will be the difference between shock impulse calculation based on integration times of 5θ as compared with 6.7θ for an underwater shock pulse?

Solution
Equation 5.1 when combined with 5.2 gives

$$I = \int_0^T p_m e^{-\frac{t}{\theta}} dt \tag{15.20}$$

If $T = 5\theta$ then $I = 0.9933 p_m \theta$ while with $T = 6.7\theta$, $I = 0.9988 p_m \theta$. The percentage difference between these two values is about 0.6% based on the 5θ value.

Example 4
Calculate the maximum bubble radius and oscillation period for a 125 kg charge of TNT detonated at a depth of (a) 80 m, (b) 15 m using Equation 3.7 or the nomogram of Figure 5.6.

If J depends on the capacity to do work as defined by the explosive power index, calculate the maximum bubble radius for a 125 kg Torpex charge detonated at 80 m depth if power indices for TNT and Torpex are 117 and 158 respectively (see Section 2.8).

Solution
With charge weight, specified depth of detonation 80 m, taking atmospheric head H_o as 10 m, Equation 5.32 gives maximum bubble radius $a_{max} = 3.91$ m and oscillation period 0.25 s. Using the nomogram gives a_{max} approximately 4 m and the same oscillation period. If detonation depth is reduced to 15 m, the effect is to increase maximum bubble radius to 6 m and oscillation time to 0.72 s. These results are for TNT. If the J value for TNT can be scaled for Torpex on the basis of the power indices then $J_{TORPEX} = 3.5 \times (158/117) = 4.73$ m$^{4/3}$ kg$^{-1/3}$ and Equation 5.32 predicts a maximum bubble radius of 5.3 m.

Example 5
How much energy is contained in the first bubble period generated by 125 kg of TNT detonated in 80 m of water? What fraction of the total explosive energy is in this bubble?

Solution
From Example 4, the maximum bubble radius was evaluated as approximately 4 m. Using Equation 5.29 with $p_o = 8.83 \times 10^5$ Pa (equivalent to a head of 90 m of water) the amount of energy in the bubble Υ is calculated to be 236.7 MJ. The total charge energy is $125 \times 4520 = 5.65 \times 10^5$ kJ or 565 MJ meaning that the bubble contains $(236.7/565) \times 100\% = 41.9\%$ of the total charge energy.

Example 6
Two 8 kg spherical charges of Comp B (containing 60% RDX and 40% TNT) are detonated, the first in air and the other underwater at a depth of 50 m. What would be the peak pressures and positive phase durations of the pulses produced by these two detonations at a range of 4 m, explaining why the values obtained are different?

Solution
Using the information of Table 3.2, an 8 kg charge of Comp B is equivalent to 9.184 kg TNT. With a range of 4 m, the corresponding scaled distance $Z(= R/W^{1/3})$ is 1.91 m/kg$^{1/3}$. From Figure 3.8 p_s is read as 1.7×10^5 Pa and positive phase duration is found to be 3.83 ms. Using the nomogram of Figure 5.3, p_m is obtained as 2.4×10^7 Pa with a time constant θ of 0.2 ms indicating a positive phase duration ($= 6.7\theta$) of 1.34 ms.

The differences between the two situations can be explained principally in terms of the compressibility of the air as compared with that of water: energy is expended in compressing air which would otherwise be available for generating a higher wave-front pressure level.

15.5   Structural response

Example 1
A lightly damped single degree of freedom system is of mass 300 kg. After 12 cycles the amplitude of free vibration reduces to 1/20th of its initial value. If the period of oscillation is 2 seconds find the logarithmic decrement of the oscillation, the damping coefficient c, the spring stiffness K and the damping factor ζ.

Solution
The solution to this problem of damped free vibration is provided by Equation 8.13. When $t = 0$ the displacement $x(0)$ is C_2 and when $t = 24$ s (corresponding to 12 cycles of vibration each of 2 s) the displacement $x(24)$ is given by

$$x(24) = e^{-24\omega\zeta}[C_1 \sin 24\omega_d + C_2 \cos 24\omega_d] \tag{15.21}$$

where ω_d is as defined in Equation 8.14. The period of damped vibration is 2 s which equals $2\pi/\omega_d$, so $\omega_d = \pi$. If this is substituted into Equation 15.21 then

$$x(24) = C_2 e^{-24\omega\zeta} = C_2 e^{-24\pi\zeta} \tag{15.22}$$

if the structure is assumed to be lightly damped and $\omega = \omega_d$. Since the amplitude after 24 seconds is 20 times less than at $t = 0$

$$\frac{x(0)}{x(24)} = \frac{20}{1} = \frac{C_2}{C_2 e^{-24\pi\zeta}} = e^{24\pi\zeta} \tag{15.23}$$

from which $\pi\zeta$ is found to be 0.125 and hence ξ is 0.0398.

From the definition of ζ (Equation 8.10) and with $\omega = \pi$, the damping coefficient is found to be 75 Ns/m. Since $\omega^2 = K/M$, the stiffness of the structure K is $300\pi^2(= 2960 \text{ N/m})$. Finally, the logarithmic decrement (defined as the natural logarithm of the ratio of amplitude at time t to the amplitude one period later) is evaluated by calculating the amplitude after one cycle $x(t_d)$ as $C_2 e^{-2\pi\zeta}$ and comparing it with the initial amplitude $x(0)(= C_2)$ giving the logarithmic decrement L as

$$L = \ln(e^{2\pi\zeta}) = 2\pi\zeta = 0.25 \tag{15.24}$$

Example 2
Compare the natural frequencies of vibration of two structures of modern construction. The first is a three storey concrete frame building of height approximately 12 m and side about 25 m. The second is a tower block of 20 storeys, approximately 80 m tall with a plan dimension of 50 m.

Solution
Two empirical relationships are quoted in Reference 15, Chapter 3, for the natural period of vibration, T of buildings of modern construction. The first relates T to building breadth B and height H (the face of area BH being at least partially loaded by the blast) of the form

$$T = \frac{0.09H}{B^{1/2}} \quad \text{seconds} \tag{15.25}$$

where H and B are in metres. The second is

$$T = \frac{H}{50} \quad \text{seconds} \tag{15.26}$$

Using the dimensions of the first building the two equations predict vibration periods of 0.216 s and 0.240 s respectively while for the second structure T is 1.02 s and 1.6 s respectively indicating the longer periods generally associated with taller structures.

Example 3

A cantilever beam is 1 m long and of rectangular section 50 mm wide by 20 mm deep. It is made of material with Young's Modulus 200 kN/mm^2 and density 8000 kg/m^3. If the natural frequency of the structure is approximately 10 Hz, show that charges of 2.3 kg and 23 kt of TNT producing a reflected overpressure (p_r) of 10^5 Pa will produce impulsive response and quasi-static response respectively. Show also that the displacement under quasi-static loading is greater than that produced impulsively by a factor of approximately 11 if a sinusoidal deflected shape is assumed.

Solution

From the scaled distance graphs of Figure 3.8, the positive phase duration of a blast wave with p_r of 10^5 Pa is found as approximately 5 ms for the 23 kg charge and about 1 s for the 23 kt device. Using the guidelines of Equation 9.10 the product of natural frequency and positive phase duration for the smaller charge is about 0.3 making this loading in the impulsive regime, while for the larger device the product is about 60 indicating quasi-static loading.

Using now the impulsive solution for deflection of this structure with the specified deflected shape (Equation 10.14) gives, on substitution, a non-dimensionalised maximum displacement W_o/L of 0.0164, while for quasi-static behaviour Equation 10.22 gives W_o/L equal to 0.179 indicating that quasi-static displacement is approximately $0.179/0.0164 = 10.9 \approx 11$ times greater than the impulsive deflection.

Example 4

A cantilever beam of length L and mass m per unit length is subjected to a uniformly distributed blast pulse producing a pressure corresponding to a load of p per unit length. Assuming the beam responds plastically with the formation of a single hinge at the root, show that the load-mass factor for the structure is 2/3.

Solution

The deflection x at a distance y from the root of the cantilever of length L can be described by the equation

$$x = x_{max} \frac{y}{L} \tag{15.27}$$

where x_{max} is the deflection of the cantilever tip. The total work done WD by a (quasi-statically) applied uniformly distributed blast load p is given by

$$WD = \int_0^L p \, x \, dy = \frac{P x_{max}}{2} \tag{15.28}$$

where P is pL. If this is equated to the work done on an equivalent system, $P_e x_{max}$, by an equivalent point load P_e, the load factor K_L is found as $\frac{1}{2}$.

The kinetic energy KE imparted to the structure in the impulsive regime is evaluated as

$$KE = \int_0^L \frac{1}{2} m \dot{x}^2 dy = \frac{1}{6} M \dot{x}^2 \tag{15.29}$$

where $M = mL$. If this is equated to the kinetic energy of the equivalent system ($\frac{1}{2} M_e \dot{x}^2$), the mass factor K_M, is found as $\frac{1}{3}$ and the load-mass factor K_{LM} is $\frac{2}{3}$.

Example 5

An elastic structure of mass M and size characterised by length L is acted upon by a blast wave of peak reflected pressure P. If the resistance of the structure is characterised by a spring of stiffness K and the blast wave has positive phase duration t_d using the techniques of dimensional analysis presented in Section 15.2, the maximum displacement of the structure x_M may be given by

$$\frac{x_M}{L} = f\left[\frac{M}{K t_d^2}, \frac{PL}{K}\right] \tag{15.30}$$

To assess the validity of the modelling process a real target and a model are to be tested in identical blast environments. The full size target has a mass of 500 kg, is 5 m long and offers a resistance of 50 kN/m. The target will displace 0.25 m in the environment of the trial. What should the mass of a 1/5th scale model of the target be, what should the resistance value be and how much would the model be expected to displace?

Solution

Using the suffixes f for full-size and m for model size, a valid trial will result if the three Pi-groups are matched simultaneously giving, after some rearrangement, the following:

$$M_m = M_f \left[\frac{K_m}{K_f}\right] \left[\frac{t_{dm}}{t_{df}}\right]^2$$

$$K_m = K_f \left[\frac{P_m}{P_f}\right] \left[\frac{L_m}{L_f}\right]$$

$$X_m = X_f \left[\frac{L_f}{L_f}\right] \tag{15.31}$$

Substitution of the data, acknowledging that the blast loads and positive phase durations will be the same for both situations, gives the model mass as 100 kg, the model stiffness as 10 kN/m and the resulting model deflection as 0.05 m.

Example 6

A series of n columns supports the flat roof slab of a structure. Each column may be taken as pin-ended of length L and carrying an equal fraction of the roof slab mass m. A bomb detonated at some height of burst above the slab produces a uniform reflected overpressure impulse i_r on an area A for each column.

By assuming that such loading may be taken as impulsive with the column deflecting in the form of a sine wave, such deformation being due to bending effects alone it can be shown, using the technique of Chapter 10, that impulse is related to the maximum central deflection of the column W_0 by the equation

$$i_r = \frac{\pi^2 W_0}{LA} \sqrt{\frac{EIm}{2Ln}} \tag{15.32}$$

where EI is the flexural rigidity of the column.

A steel universal column of Young's modulus $200 \, \text{kN/mm}^2$ and yield strength $240 \, \text{N/mm}^2$ has flange width $254 \, \text{mm}$ and a smaller second moment of area of $3.873 \times 10^7 \, \text{mm}^4$. If 20 such columns support a roof slab of total mass $700 \, \text{t}$ with each column supporting a slab area of $40 \, \text{m}^2$, will a reflected impulse of $2500 \, \text{Pa-s}$ cause the columns to buckle if each is $4 \, \text{m}$ long?

Solution

Yielding will occur at the centre of the column first where the bending moment is a maximum. If the deflected shape is a sine wave of equation

$$W = W_0 \sin \frac{\pi x}{L} \tag{15.33}$$

the maximum bending moment M_{max} (at $x = L/2$) is found as

$$M_{max} = \frac{\pi^2 EIW_0}{L^2} \tag{15.34}$$

For a beam of depth h made of material with yield strength σ_y the maximum bending moment at yield from Equation 10.15 is

$$M_{max} = \frac{2\sigma_y I}{h} \tag{15.35}$$

If these two last equations are combined, maximum displacement W_0 is given by

$$W_0 = \frac{2\sigma_y L^2}{\pi^2 Eh} \tag{15.36}$$

If this is substituted into Equation 15.32 we obtain

$$i_r = \left[\frac{2LIm}{En} \right]^{1/2} \frac{\sigma_y}{Ah} \tag{15.37}$$

which, when the data given are substituted, gives a reflected specific impulse to cause yield of $1739 \, \text{Pa-s}$. Thus, the delivered impulse of $2500 \, \text{Pa-s}$ will cause yielding of the structure.

15.6 Stress waves

Example 1 – Elastic waves
A cylindrical steel bar, of radius 100 mm, is joined at its end to two aluminium bars of radius 60 mm and 40 mm. A compressive wavefront of intensity 100 N/mm² travels along the steel bar and meets the interface between the steel and aluminium bars. Evaluate the transmitted and reflected stresses in each of the three bars.

	E kN/mm²	ρ kg/m³
Steel	210	7850
Aluminium	70	2700

Solution

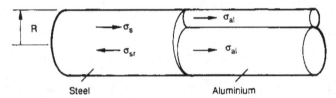

Figure 15.2 *Compound bar configuration*

The compatibility condition requires that the particle velocity either side of the interface must be the same and, therefore, that the particle velocity in both aluminium bars must be identical. Since the particle velocity is given by $\frac{\sigma}{\sqrt{E\rho}}$, the stress transmitted into both aluminium bars will be the same. Let this be σ_{al}. The compatibility condition becomes

$$\frac{\sigma_s}{\rho_s c_s} - \frac{\sigma_{sr}}{\rho_s c_s} = \frac{\sigma_{al}}{\rho_{al} c_{al}}$$

i.e. $\sigma_s - \sigma_{sr} = k \cdot \sigma_{al}$

where $k = \frac{\rho_s c_s}{\rho_{al} c_{al}} = 2.95$ (15.38)

where σ_s and σ_{sr} are incident and reflected stress levels in the steel bar. The equilibrium condition gives

$$(\sigma_s + \sigma_{sr})\pi R^2 = \sigma_{al}\pi R^2 \left(\frac{4}{25} + \frac{9}{25}\right)$$

i.e. $\sigma_s + \sigma_{sr} = 0.52\sigma_{al}$ (15.39)

Solving Equations 15.38 and 15.39 gives
$2\sigma_s = (k + 0.52)\sigma_{al}$ (15.40)

Substituting $\sigma_s = 100\,\text{N/mm}^2$ and $k = 2.95$ gives

$$\sigma_{al} = 57.6\,\text{N/mm}^2 \quad \text{and} \quad \sigma_{sr} = -69.92\,\text{N/mm}^2$$

A compressive stress of 57.6 N/mm² is transmitted into each of the aluminium bars and a tensile stress of 69.92 N/mm² is reflected back up the steel bar.

Example 2 – Plastic waves and x,t diagrams

A steel bar of length L is suddenly subjected to a compressive stress, at one end, of an intensity equal to four times the yield stress of the material of the bar. The stress is maintained for a time $T = \frac{L}{c}$, where c is the elastic wave velocity in the bar. Assuming that the modulus of deformation after yield S is a constant and equal to one-ninth of the elastic modulus E of the material, draw an x,t diagram for the period $t = 0$ to $t = 5/2T$.

Solution

An elastic stress wavefront of intensity σ_y, the yield stress of the material, propagates down the bar at a speed c. Since $\sqrt{\frac{E}{S}} = 3$, the plastic wave velocity is one-third of the elastic wave velocity. Thus a plastic wavefront of intensity $3\sigma_y$ follows the elastic wavefront at a speed $^c/_3$. At time $t = L/c$, the elastic wavefront reaches the far end of

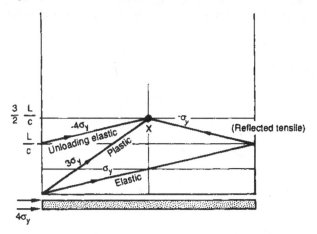

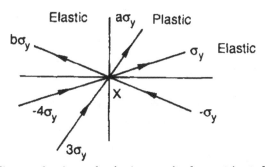

Figure 15.3 x, t *diagram showing early plastic wave development in steel bar*

the bar and simultaneously the compressive stress is removed from the near end. This is represented by an elastic unloading (tensile) wavefront of intensity $4\sigma_y$. After a time $\frac{3}{2}\left[\frac{L}{c}\right]$ the reflected tensile wavefront, the plastic wavefront and the unloading wavefront all meet at the midpoint of the bar, represented by the point X in Figure 15.3a.

We assume that both an elastic wavefront of intensity σ_y and a plastic wavefront of intensity $a.\sigma_y$ develop to the right of the midpoint and an elastic wavefront of intensity $b.\sigma_y$ develops to the left of the midpoint. The wavefronts arriving and leaving the point X are shown in Figure 15.3b. The condition of equilibrium gives

$$3\sigma_y - 4\sigma_y + b\sigma_y = -\sigma_y + \sigma_y + a\sigma_y \tag{15.41}$$

The compatibility condition gives

$$9\sigma_y - 4\sigma_y - b\sigma_y = 3a\sigma_y + 2\sigma_y \tag{15.42}$$

Solving Equations 15.41 and 15.42 gives

$$a = \tfrac{1}{2}, \quad b = \tfrac{3}{2}.$$

At first sight this answer seems impossible, since the intensity of the elastic wave $b\sigma_y$, is $1.5\sigma_y$ which is more than the yield stress. However, as will be seen from Figure 15.4, the material in that part of the bar has previously been loaded to $4\sigma_y$ (point 2) and unloaded (point 3). Reloading to a stress of $\frac{3}{2}\sigma_y$ is therefore elastic (point 4A).

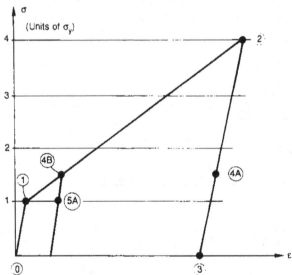

Figure 15.4   *Stress states in plastically deformed bar*

The next interaction occurs at the point Y (Figure 15.5). The equilibrium condition gives

$$\tfrac{1}{2}\sigma_y + c\sigma_y = d\sigma_y - \sigma_y \tag{15.43}$$

The compatibility condition gives

$$\tfrac{3}{2}\sigma_y - c\sigma_y = d\sigma_y + \sigma_y \tag{15.44}$$

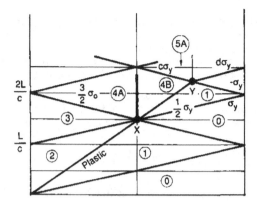

Figure 15.5 x, t *diagram showing subsequent plastic wave development in steel bar*

Solving Equations 15.43 and 15.44 gives

$$c = -\tfrac{1}{2}, \quad d = 1$$

Figure 15.4 indicates the stress and strain states in the various regions identified in Figure 15.5.

15.7 Ballistic penetration

Example 1
Distinguish between the mechanisms of 'hydrodynamic' and 'subhydrodynamic' ballistic penetration. Compare the areal density of mild steel and aluminium alloy to provide protection against the following projectile fired at:

(a) 300 m/s
(b) 3000 m/s

Projectile characteristics:

Material – hardened steel
Geometry – cylindrical with $^L/_D$ = 8 and diameter 8 mm

Materials data

Material	Hardness BHN	Density ρ kg/m³	Bulk modulus K N/m²	Young's modulus E N/m²	Yield strength σ_y N/m²	Shear strength τ_s N/m²
Mild steel	150	7830	158×10^9	206×10^9	309×10^6	275×10^6
Al alloy	75	2765	69×10^9	71×10^9	227×10^6	185×10^6

Solution
High velocity impacts involving projectiles with a high $^L/_D$ ratio cause shear stresses which are many orders of magnitude greater than the shear strength of target or projectile. The process is governed by fluid flow and is referred to as hydrodynamic penetration. Low velocity impacts of projectiles with low $^L/_D$ ratios are governed by conventional strength of material parameters and are known as subhydrodynamic.

(a)   At 300 m/s, the impact is subhydrodynamic, for which the Recht equation provides a good estimate of penetration. For the projectile, $D = 10$ mm, $L = 80$ mm, mass $= 0.049$ kg and $A_p = 78.54$ mm^2. The Recht equation for cylindrical projectiles simplifies to

$$\text{penetration} = \frac{1.61M}{bA_p}\left[V_0 - \frac{a}{b}\ln\left(\frac{a + bV_0}{a}\right)\right]$$

where the symbols have the same meaning as in Chapter 12.

For mild steel, $a = 1.978 \times 10^9$, $b = 8.793 \times 10^6$
For aluminium alloy, $a = 1.191 \times 10^9$, $b = 3.453 \times 10^6$.

So for subhydrodynamic attack of steel, penetration $= 12.51$ mm and for subhydrodynamic attack of aluminium alloy, penetration $= 24.5$ mm. So

$$\frac{(\text{Areal Density})_{\text{steel}}}{(\text{Areal Density})_{\text{al}}} = \frac{0.0125 \times 7830}{0.0245 \times 2765} = 1.44$$

(b)   At 3000 m/s, the impact is hydrodynamic and

$$\text{penetration} = L\sqrt{\frac{\rho_p}{\rho_t}}$$

So penetration into steel $= L$ and penetration into aluminium) $= L\sqrt{\frac{\rho_s}{\rho_{al}}}$ So

$$\frac{(\text{Areal density})_{\text{steel}}}{(\text{Areal density})_{\text{al}}} = \frac{\rho_s \cdot L}{\sqrt{\rho_s\rho_{al}} \cdot L} = \sqrt{\frac{\rho_s}{\rho_{al}}}$$
$$= 1.68$$

15.8 Buried structures

Example 1
Model tests on a proposed buried structure (Figure 15.6a) indicate that collapse will ultimately occur by the mechanism indicated in Figure 15.6b. The structure will be constructed in soil of bulk unit weight 18 kN/m^3 and short-term shear strength 40 kN/m^2. Determine a suitable value for M_p, the fully plastic moment per metre run of the culvert section, to provide a safety factor of 1.5 against collapse when the ground surface is subject to a uniform pressure of 150 kN/m^2.

Solution
Consider 1 metre run of the structure:

Rate of work done by external loads on section AB

$= 150.2.3\,\dot{\theta}$ kW

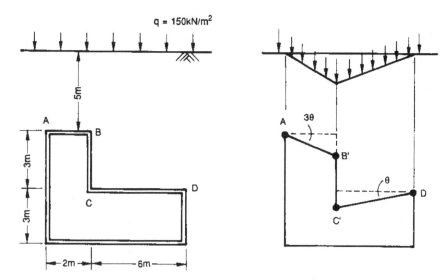

Figure 15.6 *Structure geometry and collapse mechanism for Example 1 (Section 15.8)*

Rate of work done by external loads on section CD

$$= 150.6.3\ \theta\ kW$$

Rate of work done by gravity on self-weight of soil

$$= 18(5.2.3.\ \theta + 6.8.3.\ \theta)$$

$$= \underline{3132\ \theta\ kW}$$

Rate of energy dissipation in soil

$$= 5.3.c.3\ \theta + 8.6.c.\ \theta = 93c\ \theta = 3720\ \theta\ kW$$

Rate of energy dissipation in plastic hinges $= 8M_p\theta kW$
Factor of safety

$$= 1.5 = \frac{8M_p + 3720}{3600 + 3132} => M_p = 797\ kNm/m$$

It should be noted that this is an upper band solution and it therefore gives an *unsafe* (i.e. too low) value for M_p. However, since model tests have indicated this mode of failure, the method can be legitimately adopted for design purposes.

Index

Printed in the United States
by Baker & Taylor Publisher Services